AF887576

Solar Wind Acceleration

Solar Wind Acceleration

By

Yuri V. Pisanko, Igor S. Veselovsky
and Oleg I. Yakovlev

Cambridge
Scholars
Publishing

Solar Wind Acceleration

By Yuri V. Pisanko, Igor S. Veselovsky and Oleg I. Yakovlev

This book first published 2022

Cambridge Scholars Publishing

Lady Stephenson Library, Newcastle upon Tyne, NE6 2PA, UK

British Library Cataloguing in Publication Data
A catalogue record for this book is available from the British Library

Copyright © 2022 by Yuri V. Pisanko, Igor S. Veselovsky
and Oleg I. Yakovlev

All rights for this book reserved. No part of this book may be reproduced, stored in a retrieval system, or transmitted, in any form or by any means, electronic, mechanical, photocopying, recording or otherwise, without the prior permission of the copyright owner.

ISBN (10): 1-5275-8180-2
ISBN (13): 978-1-5275-8180-7

TABLE OF CONTENTS

TABLE OF CONTENTS

PREFACE

In addition to sunshine, the Sun emits extensively into the interplanetary space of about one trillion tons of ionized hydrogen per second. The phenomenon is known as solar wind. Details of the solar wind acceleration near the Sun are poorly known. The purpose of our book is to elucidate the research of the solar wind acceleration region. Unique scientific information about the solar wind at heliocentric distances up to 3 solar radii obtained from receiving antennas directed almost to the Sun (that is the challenging technical task itself) during telecommunication sessions with “Mars 2”, “Mars 4”, “Venera 10”, “Venera 15”, “Venera 16” spacecraft and the solar wind acceleration region modelling underlie the fundamental basis for our book.

The book is based on the lectures given by Oleg I. Yakovlev for postgraduates in the Institute of Radio Engineering and Electronics of the Russian Academy of Sciences, as well as the lectures given by Igor S. Veselovsky for students at the chair of cosmic rays and space physics in Physical Department of Lomonosov Moscow State University and the lectures given by Yuri V. Pisanko for students at the chair of ocean’s thermo-hydromechanics in the Department of Aero-physics and Space

Research of Moscow Institute of Physics and Technology. The book's text is not pure lecture notes but rather materials written in preparations for lectures, sometimes not quite suitable for the oral presentation during a lecture. All three courses are directed to familiarize students and postgraduates with the circumsolar "space weather kitchen" physical principles, taking into account the author's scientific interests.

We tried to arrange the book so that each chapter could be read almost separately from the others. It allowed us to introduce the most natural notation system for each chapter: in different chapters, the same symbol could denote different physical or mathematical variables, but inside each chapter, the biunique correspondence between the symbol and the variable (physical or mathematical) remained unchanged.

We suppose that a reader knows the fundamentals of electrodynamics, magneto-hydrodynamics, and solar plasma physics, so there are many equations in the text. Not every equation is numbered to avoid unnecessarily burdening the text; numbered are only those mentioned in the text or used in deriving subsequent equations. We introduce abbreviations for an equation (Eq. (X.Y)) or a set of equations (Eqs. (X.Y)) when it is needed to mention these in the text. The first numeral in parenthesis (X) means the chapter number, the second (Y) means the number of the equation (equations) inside the chapter. When assigning a number to a system of equations, the number stands mainly across the last equation of the system. All variables and equations in the text are written in "italic type", while vectors are marked by an upper arrow in equations and are written in "bold type" in the text.

As a rule, we use the Gaussian system of units everywhere except the cases when the use of other units is dictated by established practice in the area of solar wind research: almost all researchers express the solar wind velocity in kilometres per second while magnetic flux density – in Gauss or SI unit Testa (1 nT = 10^{-5} Gauss). SI unit Hz is used as the frequency unit. In the section devoted to the electric field in the solar wind acceleration region, we use the SI system of units where the electric field is expressed in Volts per meter and the electric charge – in Coulombs. We use the off-system Angstrom unit for light wavelength and the off-system electron-volt unit for charged particle energy.

In Chapter 2, it is convenient to express the heliocentric distance in solar radii. Furthermore, the term "Astronomical Unit (AU)", which means the average distance between the Sun and the Earth, can be found in the book.

We included a few of our papers and books as references. It is stipulated by our desire to touch on subjects well known for us, which are described in detail in our listed publications. We include many references to papers published in Russian because one of our goals is to inform readers who cannot read Russian about the results published in Russian only. Indeed, our references are incomplete, and we, in advance, present our apologies to those who consider that their results are not mentioned in more detail.

We hope that our book could be helpful to young researchers and engineers specializing in solar wind and radio communications in deep space.

We are grateful to all who have read the draft or its parts, improving the text significantly. We express our sincere gratitude to Cambridge Scholars Publishing for the stimulation of writing our book.

CHAPTER 1

INTRODUCTION TO THE PROBLEM

Eclipses and the solar corona

There is no non-catastrophic phenomenon in nature that amazes the human imagination as much as solar eclipses do. Really "the death of the Sun" looked horrible for the primitive who did not understand that this was only a temporary and seemingly phenomenon. Everything begins from the solar disk distortion as if someone gnawed it from its right-hand side. The rest becomes smaller and smaller and, little by little, transforms into the sickle with spikes directed to the west. The sickle becomes as thin as a golden filament, hardly enlightening everything around, so that shadow edges look filamentary too. Eventually, the golden filament disappears, and the black spot surrounded by the mysterious halo of quickly changing rays swims on the sky at night among stars sparkling here and there. It continues for no more than 8 minutes at the equator and for no more than 6 minutes in the middle latitudes. A new golden filament breaks loose from the right-hand side of the black spot, and the light arises. The golden filament becomes thicker and transforms into the sickle with spikes directed to the east; eventually, the whole solar disk recovers.

The ancients considered an eclipse as the evil omen: “the death of the Sun”, they attributed to an unknown supernatural power. The description of the eclipse in Homer’s “The Odyssey” is related to the culmination – the day when Penelope’s suitors decided to kill Odysseus, who had come back to Ithaca. Theoclomenus rescued Odysseus when he intimidated plotters by ominous signs, including the eclipse:

> «The Sun is blotted out of the sky look there
> A lethal mist spreads all across the Earth!»
> (“The Odyssey” by Homer, Book XX, verse 356-357, translated by Robert Fagles)

The book entitled “History of the Peloponnesian War” is devoted to the war between Athens and Sparta in the 5th century B.C. The author – Athenian politician Thucydides – participated in many of described events. Eclipses often occurred during the war. The eclipse, known as the Pericles eclipse, took place in the summer of the first year of the war. Five hundred years after the events that had been described by Thucydides (the political antagonist of Pericles), Plutarch devoted XXXV paragraph of the chapter

"Pericles and Fabius Maximus" of his "Selected Biographies" to this eclipse. Pericles at the head of naval forces was ready to put to sea when the eclipse occurred, and people were frightened. Seeing helmsman's complete confusion, Pericles cloaked the helmsman and asked him whether he considered this as a misfortune. The answer was "no". Hearing this answer, Pericles asked about the difference between the two cases other than the cloak size. An adherent of the Democratic Party of Athens, Pericles meant the Moon as a cloak that hid the Sun. He knew about the Anaxagoras (500-428 B.C.) doctrine, considering the Moon as a cover hiding the Sun and the reason for darkness during an eclipse. The distance between the Moon and the Earth is four hundred times shorter than the distance between the Earth and the Sun, and its diameter is four hundred times smaller than the solar one. Therefore, the apparent diameter of both objects is almost the same: the Moon covers the Sun almost completely. During an eclipse, the Moon shadow on the Earth is the round spot of 270 km in diameter. The spot moves on the Earth surface quicker than a bullet (~ 1 km s^{-1}) along the long narrow strip where the eclipse can only be observed. On May 5, 1818, when an eclipse took place, Carl Marx was born in Trier in Germany. The eclipse was not seen in Germany but was seen in Russia, especially clear in Saint-Petersburg.

The marvelous silvery-purl beaming ring around the Sun during an eclipse is the solar corona. It is hardly ever observed during a day: it shines a million times weaker than the sunshine. Byzantine historian Leo Diaconus did the first record of the corona in his book entitled "History", written in imitation of Thucydides. He described the eclipse of 968 A.D., December 22, 4 p.m., when in calm weather the darkness covered lands and gleam stars were seen on the sky and seen was the solar disk devoid of its shine, and only its border shined faintly as if a narrow band had encircled it.

The snapshot of the corona was taken in 1851. Three components of the coronal emission were separated. The first component is K-corona (from German "kontinuirlich" – continuous), i.e. the solar light scattered on free coronal electrons in the direction of an observer. The second one is F-corona (so-called internal zodiac light), i.e. the solar light scattered on particles of the interplanetary dust slowly moving between the Earth and the Sun. The third one is E-corona, i.e. the emissions of coronal ions in spectral lines of the visible range. Mountain observations of E-corona and the lifting of spectrographs on balloons allow investigating the spectra in detail. Edlen (Edlen 1943, 30-63) identified green (5303 Å), red (6374 Å), and several other coronal spectral lines with lines of highly ionized Fe, Ni, Ca atoms. In the electron impact ionization case, the successful ionization needs the high thermal speed of a free coronal electron and, hence, the high coronal temperature. The estimates showed a temperature value of about a million degrees. Under such temperatures, the coronal gas is ionized almost entirely and is the hydrogen plasma with minor impurities of other chemical elements. Let us follow (Chapman 1957, 1-11). The coronal thermal conductivity is as follows:

$$\kappa = k_0 T_e^{5/2}$$

T_e is the electron temperature, $k_0 \approx 8\times10^{-7}$ (erg s °$K^{7/2}$) for typical coronal temperatures and electron concentrations. The heat conduction equation for the static spherically symmetric corona without any heat sources and sinks is as follows:

$$\frac{1}{r^2}\frac{d}{dr}\left(r^2 k_0 T_e^{5/2}\frac{dT_e}{dr}\right) = 0$$

Here r is the heliocentric distance.

The solution satisfying the condition $T_e \to 0$ for $r \to \infty$ is as follows:

$$T_e = T_{e0}\left(\frac{r_0}{r}\right)^{2/7} \qquad (1.1)$$

Here $T_{e0} = 10^6$ °K for $r_0 = 1{,}058\ R$, R is the solar radius.

The hydrostatic equilibrium of the corona requires:

$$\frac{dP}{dr} = -\frac{GM_s\rho}{r^2} \tag{1.2}$$

Here P is the pressure, ρ is the coronal gas mass density, G is the gravitational constant, and M_s is the solar mass. The electric neutrality of the coronal plasma implies the equality of electron and proton concentrations on the spatial scale larger than the Debye radius. Therefore,

$$\rho = n(m_e + m_p) \cong nm_p$$

Here n is the electron concentration, m_e (m_p) is the electron (proton) mass. In the case of the same proton and electron temperatures, the pressure is as follows:

$$P = 2nkT \tag{1.3}$$

Here k is the Boltzmann constant. Substituting Eq. (1.1) into Eq. (1.2), one obtains:

$$\frac{d}{dr}\left(\frac{n}{r^{2/7}}\right)=-\frac{GM_s m_p}{2\kappa T_0 r_0^{2/7}}\frac{n}{r^2}$$

The solution of this equation is as follows:

$$n(r)=n_0\left(\frac{r}{r_0}\right)^{2/7}\exp\left\{\frac{7}{5}\frac{GM_s m_p}{2kT_0 r_0}\left[\left(\frac{r_0}{r}\right)^{5/7}-1\right]\right\} \qquad (1.4)$$

Here $n_0 = n_0(r_0)$. Eq. (1.1), Eq. (1.3), and Eq. (1.4) give the pressure heliocentric distance dependence as follows:

$$P(r)=P_0\exp\left\{\frac{7}{5}\frac{GM_s m_p}{2kT_0 r_0}\left[\left(\frac{r_0}{r}\right)^{5/7}-1\right]\right\} \qquad (1.5)$$

Here $P_0 = 2n_0kT_0$. It is clear from Eq. (1.5) that the pressure monotone decreases with the heliocentric distance up to its asymptotic value at $r\to\infty$, which is:

$$P_\infty = P_0\exp\left(-\frac{7}{5}\frac{GM_s m_p}{2kT_0 r_0}\right)$$

For coronal temperatures and concentrations, P_∞ is about seven to eight orders of magnitude higher than the total pressure of the interstellar gas, the galactic magnetic field, and cosmic rays. It means that the assumption about the static corona is false.

Comets and the solar corpuscular radiation

Another enigmatic (for the ancients) phenomenon is a comet – a star with the colossal tail sometimes occupying half of the sky and appearing suddenly. Usually, a comet has not only the "tail" but also the "head". The head is the brightest top part of a comet, but the tail appears when a comet approaches the Sun. As far back as the 17th century, Kepler noted that the comet's tails were always directed away from the Sun and assumed this could be because of the sunshine. In the sixties of the 19th century, the director of Moscow astronomical observatory Bredichin (Bredichin 1861, Bd.1291) (Bredichin 1861, Bd.1305-Bd.1306) (Bredichin 1878, Bd.93) created a theory of comets. According to Bredichin, the comet's tails consist of the gas-dust mixture evaporating (due to the sunshine) from comet's heads; for sufficiently small dust particles, the solar light pressure can exceed the gravitation and comet's tails become directed from the Sun. In the eighties of the 19th century, the spectrum analysis of the light of the comet's tails revealed the ionized carbon oxide (CO^+) presence there. In the late thirties and early forties of the 20th century, comets were intensively studied in Germany. Based on observations of the Whipple-

Fedke comet in 1942, Ahnert (Ahnert 1943, 307-308) ascribed the anti-solar orientation of comet's tails to permanent corpuscular radiation of the Sun. von Hoffmeister (von Hoffmeister 1943, 265-287) also studied ionized tails of comets; the crude estimate of the solar corpuscular radiation velocity as ~ 400 km s^{-1} could be derived from his graphs. Biermann (Biermann 1951, 274-286) (Biermann 1953, 291-302) (Biermann 1957, 109-110) investigated accelerations of gas clouds (mainly CO^+) in the comet's tails. He determined the velocity and acceleration of a cloud from the sequence of comet tail snapshots representing the cloud motion along the tail. Many times he revealed accelerations of a hundred times more effective than the local gravitation. It was impossible to explain such accelerations by the solar light pressure: the latter was a few orders of magnitude smaller. Biermann attributed the discovered accelerations to the transfer of the solar corpuscular radiation momentum to the cloud in the comet tail. The carbon oxide molecule ionization he ascribed to the charge exchange reaction with solar protons and concluded that the Sun emitted corpuscles continuously in all directions.

In all fairness, let us note that Lodge (Lodge 1900, 249) wrote that terrestrial magnetic storms were caused by clouds of charged atoms or molecules moving from the Sun. FitzGerald (FitzGerald 1900, 287-288) estimated the speed of such clouds as ~ 500 km s^{-1}. Bartels (Bartels 1932, 1-52) noticed the strong tendency of moderate terrestrial magnetic storms to recur every 27 days, i.e. with the period of rotation of solar equatorial regions. The statistics of delays between the proposed solar source passage over the solar central meridian and the subsequent observation of the terrestrial magnetic storm indicated the travel time of about 3–4 days, implying the corpuscular cloud speed of about 400 km s^{-1}. In series of publications, Chapman alone (Chapman 1919, 61-83) and with Ferraro (Chapman and Ferraro 1931, 77-97) (Chapman and Ferraro 1933, 171-196) (Chapman and Ferraro 1940, 245-268) tried to explain severe magnetic storms and aurorae borealises by the action of electrically neutral clouds of electrically charged solar particles. They used sharp jumps on records of terrestrial magnetic field horizontal component variations (the so-called SSC – storm sudden commencement) as markers. The delay between the solar flare occurrence near the central meridian and the corresponding SSC on ground magnetic records was about 1–2 days, giving the cloud speed of about 1000 km s^{-1}. Forbush (Forbush 1946, 771-772) described a sudden intensity decrease of galactic cosmic rays registered with the help of ground-based ionization chambers. Usually, the ionization electric current in such devices is created by muons appearing in the terrestrial atmosphere due to atmospheric constituents' nuclear reactions with galactic and solar cosmic rays. At the end of February and beginning of March 1942, a day after the solar flare had occurred, the cosmic ray flux decreased. The phenomenon was interpreted as the submersion of the Earth into the magnetized solar plasma cloud, which prevented the penetration of galactic cosmic rays to the land. The plasma cloud speed was of the order of 1000 km s^{-1}.

Vsekhsvyatsky, Nikolsky, Ponomarev, and Cherednichenko (Всехсвятский и др. 1955, 165-176) showed that the flux of matter from the Sun should exist. Their estimate of the flux value was based on the following statements. If the corona were hydrostatic, the ratio of iron and hydrogen scales of height would be 1/56: iron ions would be observed at the coronal base only. However, iron ions radiate in the entire corona. If one assumes that the momentum of ascending flows of protons keeps the iron ions in the "suspension state", one can estimate the flux of solar protons from the observations of E-corona. Their estimates (calculated in 1955) proved to be close to later direct measurements of the solar wind proton flux on spacecraft.

The angular widening of discrete radio sources due to the diffraction and the refraction of radio waves in the near-solar plasma provided an additional research tool (Виткевич 1951, 585-588). The coronal plasma should deflect (on small angles relative to the straight line connecting the

Earth and the Sun) radio waves coming from distant cosmic sources, for example, Crab-like nebula or Taurus A. In 1955, Soviet astronomer Vitkevich (Виткевич 1955, 429-432), and his British colleague Hewish (Hewish 1955, 238-251), realized the first experimental radio-soundings of the near-solar plasma exploiting intermittent radio sources. They discovered that the widening of discrete radio sources was permanent and attributed it to the interplanetary magnetic field, which created plasma spatial inhomogeneity structures at heliocentric distances up to fifty solar radii.

Discovery of the solar wind

In 1958, Parker probably knew both Biermann's hypothesis about the solar plasma permanent outflow to the interplanetary space and Chapman's calculation according to which the backpressure of the local interstellar medium was insufficient to maintain the static solar corona. Classic studies of flows inside tubes of variable cross-sections (for example, the Laval nozzle) alluded to the possibility of the transonic gas outflow into the chamber with the vanishingly slight pressure inside – the local interstellar medium. It suggested that the transonic coronal gas outflow into the interplanetary space was a natural consequence of the high coronal temperature. To demonstrate the idea, Parker (Parker 1958, 644-675) proposed the first hydrodynamic model of the stationary solar corona, rejecting the hydrostatic corona concept. He assumed that the solar gravitation could not confine the hot coronal plasma, and the permanent stationary outflow of the coronal material into the interplanetary space took place. He figuratively named it a "solar wind". He neglected the solar rotation and assumed (for simplicity) the polytrope linkage between pressure and density in the transonic outflow. The physical process acting in both the Laval nozzle and the solar wind converts thermal chaotic motions into the directed one. In the Laval nozzle, the conversion occurs due to the nozzle face. In the solar wind, the solar gravitation limits the particles' velocities to escape. For the coronal temperature of 10^6 °K, he calculated the supersonic solar wind speed at the Earth orbit slightly higher than 450 km s^{-1}. His result was a close fit with earlier estimates based on observations of comet's tails, geomagnetic storms, and Forbush's effects. However, his model was not acknowledged at one stroke: his paper submitted to the Astrophysical Journal was rejected by two reviewers. Only the editor – Chandrasekhar (future Nobel laureate 1983), published the paper in the journal. One of the weak spots of Parker's construction was his assumption of the polytrope linkage between pressure and density. At the same time, the energy equation could be used to close the set of hydrodynamic equations. Chamberlain (Chamberlain 1961, 675-

687) attempted to fill up this gap and proposed a model based on the set of three hydrodynamic equations, namely, the continuity equation, the motion equation, and the heat transfer equation. He assumed that the heat entered into the corona due to the thermal conductivity at the very coronal base and disappeared at an infinite heliocentric distance converting entirely into the flow kinetic energy. After numerical calculations, he obtained the outflow speed at the Earth orbit of about 20 km s^{-1}. He stated that the coronal material outflow to the interplanetary space resembled a “solar breeze”.

The alternative between Parker’s solar wind concept and Chamberlain’s solar breeze concept was a matter of dispute. The final choice was given only after direct measurements of the interplanetary plasma on spacecraft. The first measurements were conducted on the “Luna 2” and the “Luna 3” spacecraft between 1959 and 1961. Plasma traps with one decelerator mesh were used. According to these measurements, the flux density of positive ions with energies higher than 50 eV per charge unit was several units multiplied by 10^8 cm^{-2} s^{-1} (Грингауз и др 1960, 1301-1304) (Gringauz 1961, 551-553). Launched in 1961, the “Explorer 10” spacecraft had plasma traps with decelerator meshes capable of measuring the flux densities of positive ions with several threshold energies per charge unit. The measured flux density was

(1-2)$\times 10^8$ cm^{-2} s^{-1} (Bonetti et al. 1963, 4061-4063), the flow velocity – of about 280 km s^{-1}, and the proton temperature – (3-8)$\times 10^5$ °K (Scherb 1964, 815-818). While direct measurements registered the plasma flux close to the expected solar wind, the observations overlapped only short intervals. Doubts about the solar wind existence were removed after three months of continuous measurements of the interplanetary plasma on the "Mariner 2" spacecraft in 1962 (Snyder and Neugebauer 1964, 89-113). A set of electrostatic analyzers allowed measuring both the proton concentration and the proton velocity. The average proton concentration proved to be 5.4 cm^{-3} and the average proton velocity – 504 km s^{-1} (Neugebauer and Snyder 1966, 4480-4484). The choice in favour of the solar wind was made owing to direct measurements on spacecraft.

Radio astronomical ground-based observations of fluctuations with frequencies in the range (0.1-10) Hz (so-called "scintillations") of fluxes from radio sources of minimal angular widths (like quasar 3C 38) (Hewish et al. 1964, 1214-1217) (Виткевич и др 1966, 55-59) revealed that the solar wind velocity increased from the equator to polar regions up to three times (Dennison and Hewish 1967, 343-344).

A phenomenon was discovered: the Sun emitted extensively into the interplanetary space of about one trillion tons of ionized hydrogen per second. In Table 1-1 presented are people contributing significantly to the discovery.

Table 1-1 Chronology of the discovery of solar wind phenomenon

High temperature of the solar corona	Edlen	1943
The anti-solar direction of the comet's tails is caused by solar corpuscular radiation.	Ahnert	1943
The Sun radiates charged particles continuously and extensively	Biermann	1951
Estimates of the solar proton outflow	Vsekhsvyatsky, Nikolsky, Ponomarev, Cherednichenko	1955
The first hydrodynamic solar wind model	Parker	1958
Direct measurements of the solar wind flux	Gringauz	1959
Solar wind direct measurements during three months	Snyder, Neugebauer	1962
Solar wind velocity estimates from observations of scintillations	Hewish, Scott, Wills; Vitkevich, Antonov, Vlasov.	1964 1966

CHAPTER 2

ACCELERATION REGION: RADIO-SOUNDING WITH SPACECRAFT SIGNALS

The pioneers in radio-sounding the near-solar plasma are radio-astronomers who exploit a quasar – the radio-source of minimal angular width. They discovered the increase of the quasar angular width when the radio ray path approached the Sun. The effect was explained by near-solar plasma inhomogeneities that led to the angular width. In 1953–1962, the phenomenon was studied in detail, and the dependence of the angular width on the heliocentric distance of the radio ray path was obtained. It allowed estimating the electron concentration and its inhomogeneities near the Sun (Erickson 1964, 1290-1311). After a few quasars had been discovered, scintillations as fast random quasar intensity variations were found out. For studying the near-solar plasma, giant antennas and special receivers for monitoring scintillations on meter wavelengths were assembled in the USSR, Great Britain, India, and the USA. This way, the near-solar plasma was investigated in the heliocentric distances between 60 solar radii and 180 solar radii. It was shown that the turbulent plasma radially moved from the Sun with the velocity of 300–400 km s^{-1} and its inhomogeneity spectrum was similar to Kolmogorov's (Лотова 1968, 292-307) (Hewish and Symonds 1969, 313-329). A method to estimate the near-solar plasma electron concentration was based on the wave group delay measurements. For pulsar NP052 situated in the ecliptic, the ray path intersects the solar corona in the fixed periods as the Earth rotates around the Sun. It allows calculating the delay and reconstructing the electron concentration heliocentric distance dependence (Hollweg 1968, 771-779). These outstanding studies demonstrated the effectiveness of the radio astronomy sounding. However, such a sounding has a critical limitation: it is based on receiving and analyzing the weak noise-type signal, which should be extracted from the solar radio emission background.

The development of deep-space radio communications opened an opportunity of sounding the near-solar plasma with highly stable monochromatic (or modulated) radio waves under the high signal-to-noise merit.

Spaceflights to Mars initiated the first deep-space radio communication experiments that showed essential group delays, strong fluctuations of frequency and phase, and the spectral linewidth due to the near-solar plasma influence. In 1967–1974, radio-sounding missions were devoted to investigating the near-solar plasma influence on trajectory measurements.

The missions revealed that all effects listed above strongly impeded deep-space communications (Goldstein 1969, 598-602) (Яковлев и др. 1974, 600-605) (Колосов и др. 1978, 555-558) (Колосов и др. 1978, 1829-1839). Spacecraft launched to planets and ground stations for deep-space radio communications allowed carrying out unique experiments on radio-sounding the near-solar plasma. In such experiments, deep-space radio communication ground stations were used as complex radio-physical installations with large parabolic antennas, high sensitive receivers, frequency standards, and precise timing, filtering, measuring and registering signals. The facilities allowed obtaining detailed information on the propagation of decimeter and centimeter radio waves in space and determining plasma parameters in the solar wind acceleration region on this basis. The Russian system of deep-space radio communications used two wavelengths λ_1 = 32 cm and λ_2 = 5.07 cm, the US one – three wavelengths λ_3 = 13 cm, λ_4 = 3.6 cm, and λ_5 = 0.96 cm. For improving the reliability of deep-space communications, two ground stations were organized in the USSR (in Evpatoria and Ussuriysk). The USA built four (in Goldstone, California, in Madrid, Spain, in Canberra and Norciya, Australia). Radio-sounding experiments were carried out using stable monochromatic high-frequency spacecraft signals during radio transmission sessions to the Earth lasting from a few minutes to many hours. Every experiment lasted 2–3 months when the ray path approached the Sun and then moved away from the Sun. Based on these radio propagation data, one may determine coronal plasma parameters at heliocentric distances from 3 to 60 solar radii. To study near-solar plasma characteristics, one uses the experimental data on radio delays and radio fluctuations and the theory of radio wave propagation in an inhomogeneous medium. Exploiting the complex experimental technique along with the statistics of signal processing and the radio wave propagation theory, one derives the information on the velocity, the electron concentration, and the turbulence in the solar wind acceleration region.

This chapter presents an overview of the heliocentric distance dependence of near-solar plasma parameters derived from radio-sounding with spacecraft signals. The heliocentric distance denoted r in the following is expressed in solar radii units in this chapter.

The described near-solar plasma parameters are related to heliolatitudes below 60° for relatively low and moderate solar activity.

Solar wind velocity

There are several means to measure solar wind velocity via radio-sounding. These are based on the analysis of radio signals received at one or two widely spaced ground stations. The first method, elaborated in (Ефимов и др 1981, 311-318) (Яковлев и др. 1980, 790-798) (Яковлев и др. 1989, 531-537), exploits the peculiarities of frequency fluctuations registered at two ground stations spaced on the significant distance BB_1. In Fig. 2-1, one can see: the solar center – O, two spaced ground stations (B and B_1), a spacecraft – A, two ray paths AB and AB_1.

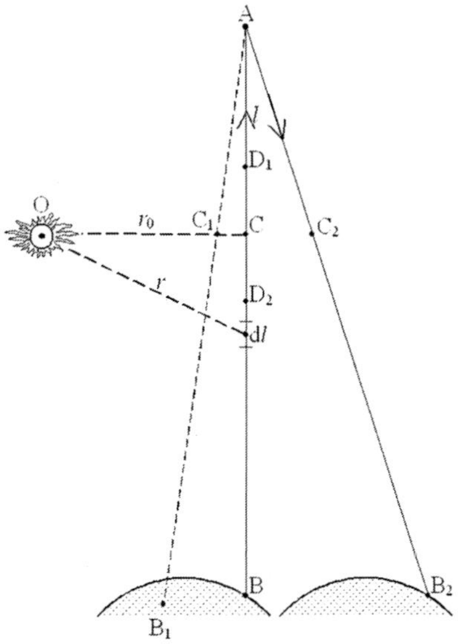

Fig. 2-1 Coronal sounding experiments under the spaced radio reception (Yakovlev and Pisanko 2018, 553)

In the region of the closest approach to the Sun defined by impact parameters $OC = r_0$ and $OC_1 = r_1$, the points C and C_1 are spaced on the distance $\Delta r = r_1 - r_0$. Therefore, the distance Δr is mainly determined by the distance BB_1 between ground stations. Radial moving solar wind plasma inhomogeneities pass through C and C_1 areas at different time moments so that similar fluctuations of frequency will be consecutively observed at the ground stations. These appear at station B_1 first and at station B next; hence, synchronous observations of frequency fluctuations will demonstrate the time delay Δt. One calculates the cross-correlation function B_f of frequency fluctuations registered at two ground stations B and B_1 and determines the time delay Δt corresponding to the maximum of the function B_f. The solar wind velocity is equal to $V = \Delta r/\Delta t$. One can precisely calculate the time delay Δt only when the function B_f has the sufficiently narrow maximum, which is the case of "frozen in" plasma inhomogeneities carried by the solar wind unchangeably through area C_1 first and area C next. Experiments revealed that the method gave satisfactory results when the cross-correlation function maximum was more significant than 0.6 and the stations B and B_1 were spaced on several thousand kilometres. The idea and the geometry of the method imply that the measured velocity corresponds to the movement of plasma inhomogeneities of several thousand kilometres in scale. It is necessary to bear in mind that radio fluctuations could be caused by plasma inhomogeneities moving with different velocities. Such a situation could be realized in both evident waves of plasma density and two almost regular solar wind streams or several plasma ejections (Efimov et al. 2003, 485-490) (Efimov et al. 2005, 2189-2194). For determining the solar wind bulk velocity *V*(*r*), it is essential to analyze the experimental correlation function of radio frequency fluctuations $B_f(\tau)$ obtained in the specified communication session for the known heliocentric distance *r*. The presence of the single narrow maximum of $B_f(\tau)$, whose value is more significant than 0.6, testifies to the single well-determined velocity value. If the diffused correlation function $B_f(\tau)$, whose value is lower than 0.6, is registered in a communication session, then the data obtained in this communication session are eliminated as doubtful. From time to time (but rarely), $B_f(\tau)$ with two well-defined maxima was registered, indicating the two-stream plasma structures.

The second method is based on the features of the operation of the space communication system in the coherent response regime. In such a regime, the spacecraft situated at point A receives the signal from the ground transmitter situated at the point B and re-transmits the signal to the ground receiver situated at the same point B (Якубов и др. 1991, 615-623). Due to the Earth orbital motion, the ground receiver appears at the point B_2 after the time $\Delta t = 2ABc^{-1}$ (*c* is the speed of light). In the coherent mode, radio waves propagate down the ray path BA first, and the ray path AB_2 next so that frequency fluctuations of radio waves will appear in area

C first and area C_2 next. In such a regime, the auto-correlation function of frequency fluctuations registered by the ground receiver has the maximum at the time moment Δt connected to the solar wind velocity V via the expression $V = CC_2/\Delta t$. The distance CC_2 is determined by the geometry and the Earth orbital motion speed. The measured solar wind velocity is supposed to reflect the motion of plasma inhomogeneities of the CC_2 scale. This method of measuring the solar wind velocity requires an uplink and a downlink. The time delay depends not only on the solar wind velocity but also on the position of the main contribution along the radio ray path in the corona (Paetzold et al., 2012, 127-152). Under quasi-stationary conditions, we assume that the main contribution to the frequency fluctuations occurs at the proximate point to the Sun on the ray path. However, examples have been reported (probably under nonstationary conditions) where this is not true.

The third method, pioneering by radio astronomers, is based on the relationship between the velocity of inhomogeneous medium transportation through the radio link and the peculiarity of the temporal spectrum of radio amplitude fluctuations recorded at a single ground station. In the theory of radio wave propagation in a statistically inhomogeneous medium (Yakovlev, 2002, 208-212), it is shown that one can determine the frequency of fluctuations F_0 as the function of the velocity V as follows:

$$V = F_0\left(\frac{2\pi\, l_1 l_2\, \lambda}{l_1 + l_2}\right)^{1/2} \tag{2.1}$$

Amplitude fluctuations are supposed to be small. The fluctuations imply that one measures several hundred kilometres' inhomogeneities (the first Fresnel zone).

Because the three methods described above encountered a difficulty for $r \leq 7$, the fourth one was elaborated for small heliocentric distances. It is based on the spectral line bandwidth change (the intensity of frequency fluctuations) when the ray path approaches to or moves away from the Sun (Efimov et al. 1977, 172-180) (Efimov et al. 2017, 1652-1662) (Ефимов и др. 1977, 322-325). According to (Ефимов и др. 1977, 322-325), the dispersion of radiofrequency fluctuations $\sigma_f{}^2$ (or the spectral linewidth square Δf^2) depends on the velocity of a plasma inhomogeneity motion through the radio link V_1 as follows:

$$\sigma_f^2 \propto V_1^{\,p-2} \tag{2.2}$$

Here p is the spectral index of the plasma turbulence spatial spectrum. The velocity V_1 in Eq. (2.2) is equal to the sum of (or the difference between) the solar wind velocity V and the velocity dr_0/dt of the ray path approach (or move away) due to the motions of the Earth and a spacecraft. Therefore,

$$\sigma_f = \left(V \pm dr_0 / dt\right)^{(p-2)/2} \tag{2.3}$$

Sign "+" corresponds to the ray path approach to the Sun; sign "–" corresponds to the motion of the ray path away from the Sun. Eq. (2.3) allows the solar wind velocity determination when V and dr_0/dt are of the same scale, and σ_f is given both for the ray path approach to the Sun and the ray path motion away from the Sun. From experimental heliocentric distance dependencies $\sigma_f(r)$ obtained in both cases, one can calculate the ratio $\beta = \sigma_{ap}/\sigma_{ma}$. In the case of $p = 3$ for $r \leq 7$, Eq. (2.3) degenerates into the equation:

$$V = \left(\frac{\beta^2 + 1}{\beta^2 - 1}\right)\frac{dr_0}{dt} \tag{2.4}$$

Eq. (2.4) gives the solar wind velocity in case of known (from the spacecraft trajectory data) dr_0/dt and measured values of β. As is easy to see, the method has some uncertainty. During the stage of ray path approach to the Sun and the stage of ray path motion away from the Sun (which are spaced in time when the spacecraft moves), r_0 can (in some moments) be of the same value, and, according to Eq. (2.4), so does the solar wind velocity. Therefore, Eq. (2.4) estimates the solar wind velocity only when the Sun is quiet and does not vary enormously during a month.

An additional method was proposed and realized to determine the solar wind velocity at small heliocentric distances. The method was based on the difference in the refraction of decimetre and centimetre radio waves (Efimov et al. 2005, 2189-2194). In the near-solar plasma, ray paths are curved so that the decimetre ray path is situated closer to the Sun. In contrast, the centimetre one is situated almost on the straight line AB (the refraction of centimetre radio waves is weak). The ray paths are separated on the distance Δr while both signals are received at the same place. To calculate the time interval Δt during which the plasma inhomogeneity moves the distance Δr, one applies the correlation of decimetre and centimetre radio wave fluctuations. The solar wind velocity is defined as $V = \Delta r/\Delta t$. The refraction is intensely varied, and one has to use model electron concentration heliocentric distance dependencies to calculate Δr keeping in mind that these dependencies are strongly variable in reality.

Several attempts to derive the solar wind velocity heliocentric distance dependence were undertaken in 1976–1997. The results are shown in Figs. 2-2 and 2-3, where different signs mark the dependencies derived by different methods. The results of the first method are marked by squares, the second – by triangles, the third – by circles, the fourth – by crosses.

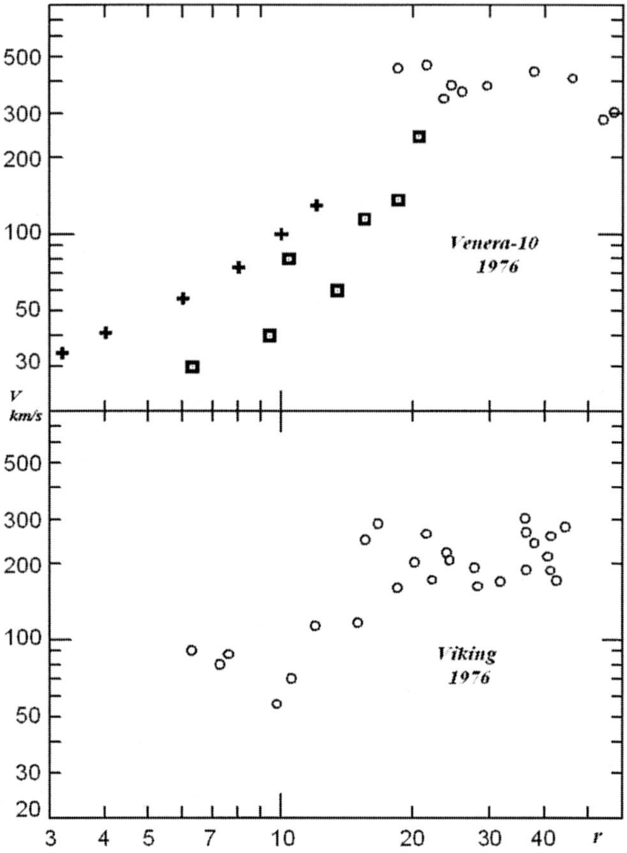

Fig. 2-2 Solar wind velocity (in km s^{-1}) experimental heliocentric distance dependencies in low solar activity; the heliocentric distance is expressed in solar radii units; the results of the first method are marked by squares, the second – by triangles, the third – by circles, the fourth – by crosses (Yakovlev and Pisanko 2018, 554)

The reliable heliocentric distance dependence in the solar wind acceleration region was obtained from radio signals of the ''Venera 10'' spacecraft (Efimov et al. 1977, 172-180) (Ефимов и др. 1981, 311-318) (Kolosov et al. 1982, 664-674) (Яковлев и др. 1980, 790-798). The dependence is shown in the upper graph of Fig. 2-2. The measurements were carried out at the wavelength λ_1 in April–July 1976 in the solar minimum. Forty-eight radio-communication sessions were chosen for the analysis, and the heliocentric distance varied from 58 to 3. To estimate the solar wind velocity, the third method was applied when $18 < r < 58$, the first – when $6 < r < 21$, the fourth – when $3 < r < 12$. The graph shows no noticeable velocity variation at heliocentric distances $30 < r < 60$: the average solar wind velocity equals 380 ± 80 km s^{-1}. The velocity decreases at heliocentric distances $6 < r < 20$. At heliocentric distances $3 < r < 5$, the average solar wind velocity equals 42 ± 15 km s^{-1}.

Another heliocentric distance dependence of solar wind velocity was obtained from radio signals of the ‘‘Viking” spacecraft (Tyler et al. 1981, 318-332) in October 1976 – January 1977 in the solar minimum. The radio signal amplitude fluctuations at the wavelength λ_3 were used, and the heliocentric distance varied from 45 to 6. The third method was applied to estimate the solar wind velocity. The dependence (Tyler et al. 1981, 318-332) is shown in the lower graph of Fig. 2-2. It is seen that the solar wind velocity varies in the interval $240 < V < 320$ km s^{-1} at heliocentric distances $24 < r < 40$, and it is equal to 74 ± 30 km s^{-1} at heliocentric distances $7 < r < 9$. The good conformity of ‘‘Venera 10” and ‘‘Viking” results at heliocentric distances $7 < r < 20$ testifies for the strong velocity increase with the heliocentric distance. The data from both missions reveal the essentially constant solar wind velocity at heliocentric distances $30 < r < 50$.

From March to September 1984, a large amount of information was obtained from the “Venera 15” and the “Venera 16" spacecraft; exploited were the decimetre wavelength λ_1 and the centimetre wavelength λ_2. All four methods mentioned above were applied to estimate the solar wind velocity. The use of many communication sessions along with two wavelengths and four methods enhances the reliability of the derived solar wind velocity heliocentric distance dependence (Ефимов и др. 1987, 2025-2032) (Ефимов и др. 1990, 581-587) (Рубцов и др. 1987, 620-625) (Яковлев и др. 1987, 251-257) (Яковлев и др. 1988, 1290-1299) (Яковлев и др. 1989, 531-537) (Якубов и др. 1991, 615-623) shown in the upper graph of Fig. 2-3.

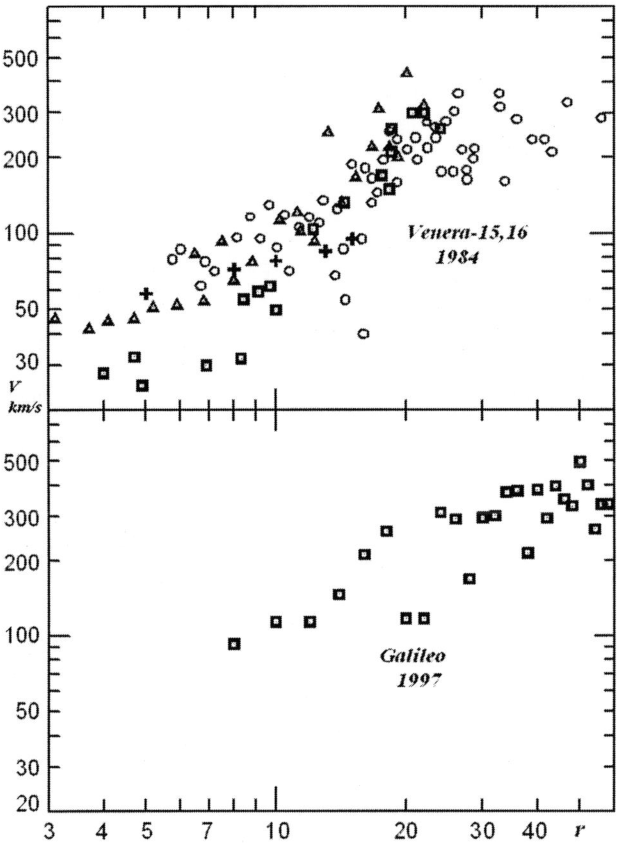

Fig. 2-3 Solar wind velocity (in km s^{-1}) experimental heliocentric distance dependencies in moderate solar activity; the results of the first method are marked by squares, the second – by triangles, the third – by circles, the fourth – by crosses (Yakovlev and Pisanko 2018, 555)

Similar to the upper graph of Fig. 2-2, there is no noticeable velocity variation at heliocentric distances $36 < r < 50$: the average solar wind velocity equals 280 ± 50 km s^{-1}. The velocity decreases to the average value of 50 ± 20 km s^{-1} at heliocentric distances $4 < r < 6$. The use of different methods to measure the solar wind velocity provides the opportunity to estimate the precision of measurements: for $r > 16$, different methods lead to the same velocity value, whereas for $r < 9$, the third (amplitude) method leads to the overestimated velocity value as against the first and the second (frequency) ones lead.

The solar wind velocity heliocentric distance dependence derived from the radio-sounding data of the ‘‘Galileo” spacecraft (at the wavelength λ_3) in January–February 1997 under moderate solar activity (Ефимов и др. 2009, 773-784) is shown in the lower graph of Fig. 2-3. The first method was applied to estimate the solar wind velocity; frequency fluctuations registered at several long spaced ground stations were analyzed. The authors noted the solar wind velocity variability: the velocity varied from 160 km s^{-1} to 600 km s^{-1} during five hours. Each point in the graph is the average over the interval ±1 solar radius (over 10–20 measurements) around the corresponding heliocentric distance. As is easy to see, there is no noticeable velocity variation at heliocentric distances $40 < r < 60$: the average solar wind velocity equals to 370 ± 40 km s^{-1}. At heliocentric distances $7 < r < 30$, the velocity decreases and its average value equals to 100 ± 30 km s^{-1} at heliocentric distances $8 < r < 10$.

Let us consider the solar wind velocity heliocentric distance dependence derived from the radio-sounding data of the ‘‘Ulysses” spacecraft (at the wavelengths λ_3 and λ_4) in August–September 1991 under high solar activity (Janardhan et al. 1999, 157-172) (Wohlmuth et al. 1997, 617-628). The first method was applied in (Janardhan et al. 1999, 157-172) to estimate the solar wind velocity, the second – in (Wohlmuth et al. 1997, 617-628). According to Janardhan et al. (1999, 157-172) and Wohlmuth et al. (1997, 617-628), there was the high dispersion of solar wind velocities at small heliocentric distances.

The correlation function analysis of frequency fluctuations registered at two spaced ground stations revealed the communication sessions when two correlation maxima existed (Chashei et al. 2005, 1454-1460). The latter authors established that two values for the velocity had taken place on the ray path during the communication sessions. This could be interpreted as the simultaneous presence of two solar wind streams with different velocities on the ray path.

The additional analysis (Efimov et al. 2003, 485-490) (Efimov et al. 2005, 2189-2194) confirmed the high variability of the solar wind velocity. Let us note that because of the "Ulysses" spacecraft trajectory, the solar wind velocities measured at heliocentric distances $r < 10$ refer to high heliolatitudes. Under high solar activity the average solar wind velocity was equal to 390 ± 100 km s^{-1} at heliocentric distances $23 < r < 34$, and 130 ± 40 km s^{-1} at heliocentric distances $7 < r < 8$.

It is difficult to determine the solar wind velocity at small heliocentric distances: in Fig. 2-2 and 2-3, one can see six experimental points at heliocentric distances 3–4 with the mean velocity value equal to 40±8 km s^{-1}. It is interesting to note that the solar wind velocity determined by the radiolocation of the Sun (from Doppler shifts) is estimated as ~ 40 km s^{-1} at heliocentric distances 2–4 (James 1970, 143-162).

Let us discuss the solar wind velocity heliocentric distance dependence at heliolatitudes below 60° for low and moderate solar activity. The primary influence of the solar wind on the radio wave propagation is concentrated in the closest to the Sun part (r_0) of the ray path because the solar wind plasma density decreases with heliocentric distance. This point permits us to consider the solar wind velocity heliocentric distance dependence ($V(r_0)$) though the solar wind presents everywhere along the ray path. To approximate the solar wind velocity heliocentric distance dependence, one can use the empirical equation:

$$V_a(r) = V_{Earth} th(x) = V_{Earth}\left(e^{x} - e^{-x}\right)\left(e^{x} + e^{-x}\right)^{-1} \qquad (2.5)$$

Here $x = \eta r$, V_{Earth} is the solar wind velocity at the Earth orbit (r = 215). For free parameter values V_{Earth} = 300 km s^{-1} and η = 0.04, Eq. (2.5) rather well describes the solar wind velocity heliocentric distance dependence at heliocentric distances $3 < r < 50$. The approximation refers to heliolatitudes below 60° in low and moderate solar activity.

According to Jensen et al. (2016, 465-485) Яковлев и др. (1987, 251-257) and Яковлев и др. (1988b, 1290-1299), the Alfvenic velocity is achieved at the heliocentric distance $16 < r_a < 18$.

It is possible also to use another empirical equation (Yakovlev and Yakovlev 2019, 237-242):

$$V_a = V_{Earth}\left\{1+\exp\left[\alpha\left(r_p - r\right)\right]\right\}^{-1} \qquad (2.6)$$

Here $V_{Earth} = 325$ km s^{-1}, $\alpha = 0.2$, $r_p = 13.7$.

Solar wind density

First estimates of the electron concentration near the Sun were derived from the Thomson scattering of the solar light on the coronal electrons. The phenomenon was observed as K-corona using coronographs. Based on the observations, the average spherically symmetric electron concentration heliocentric distance dependence was approximated by the power law:

$$N(r) = Ar^{-6} + Br^{-2}$$

Here $A = 1.3\times10^8$ cm^{-3} and $B = 1.1\times10^6$ cm^{-3} at heliocentric distances $r > 2$. Following the discovery of a pulsar situated approximately in the ecliptic, a method based on the radio-wave group delay measurement was utilized to estimate the electron concentration of the near-solar plasma (Weisberg et al. 1976, 252-258, and references therein) (Dollfus and Mouradian 1981, 3-10). One measures the difference in arrival times of pulsar impulses on two frequencies proportional to the total electron content along the ray path. Assuming the spherically symmetric electron concentration distribution, one estimates the electron concentration at several heliocentric distances with the help of the method. Troubles of the method are connected to strong interfering solar radio-frequency emissions that prevent obtaining the electron concentration heliocentric distance dependence in the solar wind acceleration region.

The first solar wind density heliocentric distance dependence derived from radio-sounding with spacecraft was published in 1977 (Muhleman et al. 1977, 943-957). The authors exploited the single frequency communication line with the ‘‘Mariner 6” and the ‘‘Mariner 7” spacecraft to measure the radio delay. The coherent response mode for modulated radio waves used in the experiments allows the precise determination of the Doppler frequency shift and the distance to the spacecraft. These data, along with celestial mechanics, allow the reconstruction of the spacecraft trajectory, including the distance between the spacecraft and the ground station (Fig. 2-1). When the ray path approaches the Sun, the apparent distance increases because of the radio delay. The delay is proportional to the double total electron content along the ray path. Radio-sounding from the ‘‘Mariner 6” and the “Mariner 7” spacecraft realized in April-May 1970 in the solar minimum provided the electron concentration heliocentric distance dependence at heliocentric distances $4 < r < 14$. The dependence was supposed as follows:

$$N(r) = Ar^{-6} + Br^{-\xi} \quad (2.7)$$

Free parameters A, B, ξ were adjusted from radio-sounding experiments. According to the ''Mariner 6'' data, $A = 0.69\times10^8$ cm^{-3}, $B = 0.54\times10^6$ cm^{-3}, $\xi = 2.05$, and $A = 1.3\times10^8$ cm^{-3}, $B = 0.66\times10^6$ cm^{-3}, $\xi = 2.08$ according to the ''Mariner 7'' data. One can estimate the electron concentration at the heliocentric distance $r = 10$ as $N = 5\times10^3$ cm^{-3} from Eq. (2.7).

More reliable information on the solar wind density heliocentric distance dependence was collected with the help of the dual-frequency communication line when trajectory data uncertainties weakly influenced the total electron content. The dual-frequency technique allows measuring the difference between radio delays $\Delta t_{3,4}$ on two frequencies (namely, t_3 and t_4) when the ray path AB approaches the Sun and moves away from the Sun. It is important to define $\Delta t_{3,4}$ as the differential group delay time, a quantity proportional to the total electron column density. It should be distinguished from the differential phase delay time, which yields only the changes in the electron column density from the start of the measurements. The experimental curve $\Delta t_{3,4}(r_0)$ determines the total electron content J along the ray path AB that allows deriving the electron concentration heliocentric distance dependence $N(r)$. According to Fig. 2-1, the impact parameter (that is the closest approach to the Sun) of the ray path CO is equal to r_0 and the distance to a current point of the ray path – r_1. On high frequencies, the plasma refraction coefficient is related to the electron concentration as $n = 1 - DNf^{-2}$; therefore, the radio-wave delay is proportional to the total electron content:

$$\Delta t = Df^{\;2}c^{\;1}\int_A^B N(r)dl = Df^{-2}c^{-1}J(r_0) \qquad (2.8)$$

As it is easy to see (from Eq. (2.8)), the difference between radio delays $\Delta t_{3,4}$ measured on frequencies f_3 and f_4 is connected to the total electron content as follows:

$$J(r_0) = \frac{\Delta t_{3,4}(r_0)c}{D\left(f_3^{-2} - f_4^{-2}\right)}$$

Here $D = 40.4$ m^3 Hz^2, c is the speed of light in vacuum (in m s^{-1}), N is the electron concentration (in m^{-3}), f is the frequency (in Hz). Taking into account that the length element dl of the ray path AB is related to the heliocentric distance r as $dl = (r^2 - r_0^2)^{-1/2} r dr$, one can write:

$$J(r_0) = 2\int_{r_0}^{CB} \frac{N(r)rdr}{\left(r^2 - r_0^2\right)^{1/2}} \qquad (2.9)$$

In Eq. (2.9), the inessential approximation CB ≈ CA is accepted, so the integral along the ray path AB is equal to the integral along the segment CB multiplied by 2 (Fig. 2-1). The total electron content is derived from the experimental data, but the goal is to determine the electron concentration heliocentric distance dependence $N(r)$. The Abel integral transformation gives the rigorous solution of the problem:

$$N(r) = -\frac{1}{\pi}\int_{r}^{CB}\left(\frac{dJ}{dr_0}\right)\frac{dr_0}{\left(r^2 - r_0^2\right)^{1/2}} \tag{2.10}$$

The limited number of experimental points (each measured with substantial errors) yields the dependence $J(r_0)$; therefore, it is difficult to find the derivation dJ/dr_0 from the experimental data. One uses an approximation of the electron concentration heliocentric distance dependence $N(r)$ and finds the approximation parameters from Eq. (2.9). For $r > 4$, a possible approximation is as follows:

$$N(r) = Br^{-\xi}F(\theta) \tag{2.11}$$

Here $F(\theta)$ is the model approximation of the electron concentration heliolatitude dependence (various according to various authors). If a spacecraft moves almost in the ecliptic (the heliolatitude is lower than 40°), one supposes $F = 1$. Eq. (2.11) has two sought parameters: B and ξ. As there is no solar wind velocity heliocentric distance dependence for $r > 30$, one accepts $N(r) \sim r^{-2}$ and $\xi = 2$. However, the solar wind velocity increases up to 5 – 7 times when the heliocentric distance increases from 7 to 25; hence, ξ will differ from 2 in this heliocentric distance interval. The parameters B and ξ were found from experimental dependencies $J(r_0)$ by fitting (Anderson et al. 1987, L141-L143) (Bird et al. 1994, 373-381) (Bird et al. 1996, 441-448) (Krisher et al. 1991, L57-L61) (Muhleman and Anderson 1981, 1093-1101) (Patzold et al. 1995, 3313-3317) (Patzold and Bird 1998, 1845-1849) (Woo 1996, L87-L90).

Dual-frequency radio-sounding experiments were realized with the "Viking" spacecraft that had the dual-frequency (f_3 and f_4) communication line (Muhleman and Anderson 1981, 1093-1101). Experiments were carried out in November–December 1976 in the solar minimum when the difference between radio delays $\Delta t_{3,4}$ was measured in the impact parameter heliocentric distance interval $4 < r_0 < 170$ for the ray path AB both approached the Sun and moved away from the Sun. According to the authors, the experimental dependence $\Delta t_{3,4}(r_0)$ was well approximated as follows:

$$\Delta t_{3,4} = 1.2 \cdot 10^5 r^{-1.7} + 2.55 \cdot 10^{-4} r^{-1}$$

Here $\Delta t_{3,4}$ was expressed in nanoseconds. On these grounds, the authors supposed the electron concentration heliocentric distance dependence approximation as follows:

$$N(r) = B_1 r^{-2.7} \exp\left(-\frac{\theta^2}{\theta_1^2}\right) + B_2 r^{-2}$$

Here θ is the heliolatitude. The exponential multiplier of the first term describes the supposed heliolatitudinal decrease of the electron concentration; under $\theta_1 = 8^o$ and for $|\theta| < 40^o$, it can be taken as unity. The best correspondence to the experimental data gives $B_1 = 1.32\times10^6$ cm^{-3} and $B_2 = 2.3\times10^5$ cm^{-3}. Another heliocentric distance dependence of the solar wind density was obtained with the help of Eq. (2.10). One calculated the derivative dJ/dr_0 numerically after averaging the experimental data $J(r_0)$ using the running window of the width $\Delta r = 2$. Both methods lead to the same result that testifies for the calculated dependence $N(r)$. According to the data, $N = 6\times10^3$ cm^{-3} at $r = 10$ and $N = 10^3$ cm^{-3} at $r = 20$. The data quite well corresponds to the dependence $N(r) \sim r^{-2}$ for $r > 40$. The systematic excess of the data over the dependence $N(r) \sim r^{-2}$ is evident for $r < 20$ and testifies for the solar wind velocity decrease when the heliocentric distance decreases.

The results of measurements of $\Delta t_{3,4}$ and $J(r_0)$ from the near-solar plasma radio-sounding with the ‘‘Voyager 2” spacecraft in November–December 1985 in the solar minimum are presented in (Anderson et al. 1987, L141-L143). The spacecraft was located near the ecliptic such that the ray path to the Earth remained below 8° at the proximate point to the Sun. The limited number of $\Delta t_{3,4}$ measurements in the impact parameter interval $6 < r_0 < 38$ were realized in the experiments. To derive the dependence $N(r)$, the authors exploited Eq. (2.11) for $F(\theta) = 1$ and found the values of parameters B and ξ. It turned out that a strong asymmetry was observed for the solar wind density: for the ray path approach to the Sun, it was found $\xi = 2.6$, $N(10) = 6\times10^3$ cm^{-3}, whereas $\xi = 2$, $N(10) = 9.5\times10^3$ cm^{-3} when the ray path moved away from the Sun. It was shown (Woo 1996, L87-L90) with the help of coronograph observations that the asymmetry (the disparity in radial profiles) was caused by longitudinal variations stemming from probing of significantly different source regions.

One more dual-frequency radio-sounding was realized with the ‘‘Voyager 2” spacecraft in December 1988 in the solar maximum (Krisher et al. 1991, L57-L61). The majority of delay measurements was conducted in the impact parameter interval $10 < r_0 < 87$ at heliolatitudes below 20º. After data processing based on Eq. (2.11), two pairs of parameter values were found, namely, $\xi = 2.08 \pm 0.05$, $B = (2.93 \pm 0.07)\times10^6$ cm^{-3} and $\xi = 2.28 \pm 0.05$, $B = (7.08 \pm 0.24)\times10^6$ cm^{-3}. Despite the lack of data for $r < 10$, the estimates $\xi = 2.7 \pm 0.2$, $B = (2.1 \pm 0.4)\times10^7$ cm^{-3} at short heliocentric distances were derived. According to the experiments, the values $N(10) = 3.4\times10^4$ cm^{-3} and $N(20) = 5.8\times10^3$ cm^{-3} testify for the substantial enhancement of the electron concentration in the solar maximum.

Radio-sounding with the ‘‘Ulysses” spacecraft was realized in May–September 1991 in the solar maximum (Bird et al. 1994, 373-381) (Patzold et al. 1995, 3313-3317). Fifty-six measurements of the delay $\Delta t_{3,4}$ were conducted in the impact parameter interval r_0 = (40–5) at heliolatitudes below 20°. The total electron content at high heliolatitudes (where the heliolatitudinal dependence $F(\theta)$ should be considered) was determined in six experiments. It was found $\xi = 2.54 \pm 0.05$, $B = (3.61 \pm 0.04)\times10^6$ cm^{-3} based on Eq. (2.11) for the ray path approach to the Sun, whereas $\xi = 2.42 \pm 0.05$, $B = (2.26 \pm 0.03)\times10^6$ cm^{-3} when the ray path moved away from the Sun; $N(20) = (1.7 \pm 0.1)\times10^3$ cm^{-3}. The authors noted irregular deviations from Eq. (2.11) and connected the deviations with active solar regions.

One more radio-sounding with the “Ulysses” spacecraft was realized in February–March 1995 (Bird et al. 1996, 441-448) (Patzold and Bird 1998, 1845-1849). The near-solar plasma was sounded in the impact parameter interval $21 < r_0 < 30$ when the heliolatitude varied from $\theta = 30°$ to $\theta = 84°$. The electron concentration heliolatitude dependence was derived in (Bird et al. 1994, 373-381). Models from earlier work (Tyler et al. 1981, 318-322) (Muhleman and Anderson 1981, 1093-1101) were poor representations of the latitude dependence.

Multiple measurements of electron concentration provided solar wind density heliocentric distance dependencies at heliolatitudes below 40° under various solar activity levels. These do not demonstrate any clear solar activity dependence of the average electron concentration. One can accept $N(20) = 1.4\times10^3$ cm^{-3} and $N(10) = 6.7\times10^3$ cm^{-3} for the average electron concentration. The radial falloff exponent varies from 2.05 to 2.6; the closer is the ray path to the Sun, the greater is the radial falloff exponent. The comparison of electron concentration heliocentric distance dependencies derived from both radio-sounding and Thomson's scattering demonstrated the concordance of data sets obtained by different methods in the heliocentric distance interval $3 < r < 6$. The detailed comparison between heliocentric distance dependencies $N(r)$ obtained by radio-sounding and Thomson's scattering is given in (Yakovlev, 2002, 208-212) (Jensen et al. 2016, 465-485). The comparison shows the good correspondence between the two techniques for $3 < r < 5$, but the distinction occurs for $6 < r < 25$. It is natural because the Allen-Baumbach traditional models worse describe experimental data of radio-sounding.

The flow through the solar centered sphere:

$$L = 4\pi r^2 VN = 4\pi (215)^2 V_{Earth} N_{Earth},$$

(N_{Earth} is the electron concentration at the Earth orbit) along with Eq. (2.5) gives:

$$N_a = N_{Earth}(215)^2\left[r^2 th(x)\right]^{-1} \tag{2.12}$$

Using Eq. (2.6) instead of Eq. (2.5) gives:

$$N_a(r) = \frac{N_{Earth}(215)^2}{r^2}\left\{1+\exp\left[\alpha\left(r_p - r\right)\right]\right\} \tag{2.13}$$

Eqs. (2.12) and (2.13) are valid for constant L and no heliolatitude dependencies of solar wind velocity and electron concentration. According to Eqs (2.12) and (2.13), the electron concentration decreases with the heliocentric distance as $N(r) \sim r^{-2}$ for $r > 40$ where the solar wind velocity does not depend on the heliocentric distance. Electron concentrations calculated from Eq. (2.12) under the parameters $N_m = 8$ cm^{-3} and $\eta = 0.04$ agree with experimental data under low solar activity. Due to the solar wind changeability, Eq. (2.12) gives approximate N: $N(10)$ variates within 30% under moderate solar activity. Still, in coronal mass ejections, $N(10)$ sometimes enhances up to three times.

Power and acceleration

Let us calculate the power and the acceleration of the solar wind from Eqs. (2.5), (2.12), and (2.13) (Yakovlev and Pisanko 2018, 552-566). The solar mass loss due to the solar wind is as follows:

$$\frac{dM_s}{dt} = 4\pi r^2 m_p V(r) N(r) \qquad (2.14)$$

Here m_p is the proton mass. Radio-sounding allows to estimate the average solar wind velocity as V = 100 km s^{-1} and the average electron concentration as $N = 7\times10^3$ cm^{-3} at the heliocentric distance r = 10. Hence, one calculates $dM_s/dt = 0.7\times10^9$ kg s^{-1} from Eq. (2.14). At the same time, Eq. (2.14) gives $dM_s/dt = 1.1\times10^9$ kg s ^{1} at the Earth orbit (r = 215) in case of V = 300 km s^{-1} and N = 8 cm^{-3}. The total power of the solar wind is as follows:

$$W = \frac{dM_s}{dt} \cdot \frac{V^2}{2} \qquad (2.15)$$

The substitution of Eq. (2.5) and Eq. (2.14) into Eq. (2.15) gives:

$$W = \frac{dM_s}{dt} \cdot \frac{V_{Earth}^2}{2} [th(x)]^2 \qquad (2.16)$$

The solar wind power heliocentric distance dependence plotted from Eq. (2.16) is shown in Fig. 2-4 as number 1.

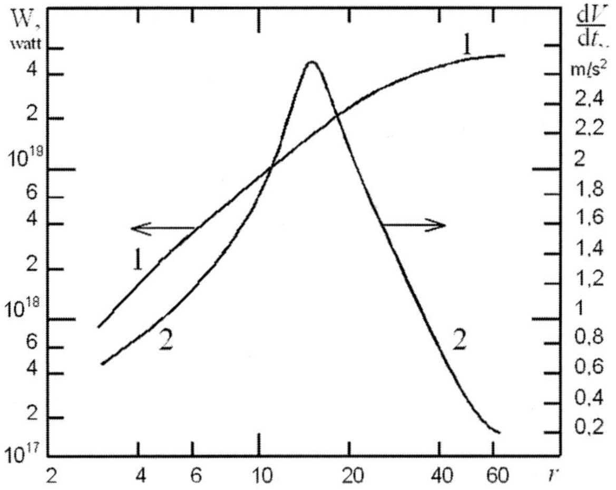

Fig. 2-4 Power *W* and acceleration *dV/dt* in the solar wind acceleration region (Yakovlev and Pisanko 2018, 559)

As is easy to see, solar wind power is enhanced approximately thirty times when the heliocentric distance increases from 6 to 30. The solar wind acceleration is as follows:

$$\frac{dV}{dt} = \frac{dV}{dr}\frac{dr}{dt} = \frac{dV}{dr}V$$

From Eq. (2.5), one calculates:

$$\frac{dV}{dt} = 4\eta \frac{V_{Earth}^2}{R} \frac{th(x)}{\left(e^x + e^{-x}\right)^2}$$

Here R is the solar radius, $x = \eta r$, $4\eta V_{Earth}{}^2 / R = 20.6$ m s^{-2} for $\eta = 0.04$, $V_{Earth} = 300$ km s^{-1}. The solar wind acceleration heliocentric distance dependence is shown in Fig. 2-4 as number 2. As is easy to see, the maximum acceleration (2.7 m s^{-2}) achieves at $r = 15$.

$V(r)$ and $N(r)$ substantial changes lead to uncertainties of averaged approximations, represented by Eq. (2.5) and Eq. (2.12).

Temporal spectra of frequency and amplitude fluctuations

It is possible to derive statistical characteristics of irregularities of the solar wind plasma from temporal spectra parameters of electron concentration fluctuations, which, in turn, can be determined from measurements of temporal spectra of radiofrequency and amplitude fluctuations. Let us consider the plasma turbulence characteristics derived from the analysis of radio wave fluctuations in the solar wind acceleration region between 3 and 40 solar radii.

One can determine the amplitude fluctuation spectrum $G_a(F)$ only within narrow limits. For the fluctuation frequency $F > 20$ Hz, the spectral intensity G_a is usually tiny, and it is difficult to separate it from noise fluctuations. For $F < 0.3$ Hz, it is challenging to determine the spectral intensity G_a because of the enhancement of variations of the signal receiver system. It has been possible to obtain plausible amplitude spectra for the frequency interval between 0.3 Hz and 20 Hz. A typical spectrum G_a is shown in Fig.2-5A.

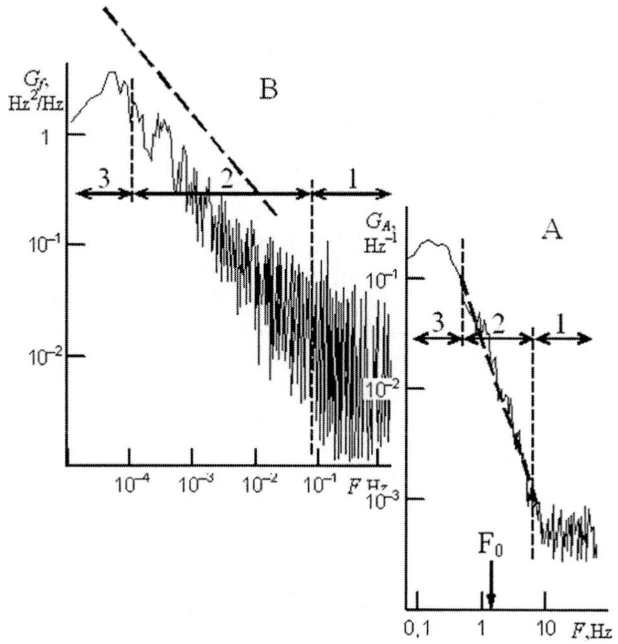

Fig. 2-5 A typical amplitude fluctuation spectrum (A) and a spectrum of carrier fluctuations (B)

For G_a, seen is the wobble (1) as well as the regular increase (2) and (for low frequencies) the almost constant plateau (3). In the small range of fluctuation frequencies F, the spectrum of amplitude fluctuations can be expressed as the power-law dependence $G_a \propto F^{-\chi a}$, where χa is the spectral index of the amplitude fluctuation spectrum. The dashed line shows this approximation in Fig. 2-5A. The arrow points to the characteristic frequency F_0 used in Eq. (2.1) to determine the solar wind velocity (V).

Also, it is possible to find the spectrum of frequency fluctuations. An example of the spectrum of carrier fluctuations (G_f) is shown in Fig.2-5B. On the broader interval (2) of fluctuation frequencies F, the spectrum density $G_f(F)$ can also be expressed as the power-law dependence $G_f \propto F^{-\chi f}$, where χf is the spectral index of the temporal spectrum of frequency fluctuations. For very low frequencies of fluctuations, experimental spectra become plane or demonstrate a weak maximum (3). Many spectra $G_f(F)$ and $G_a(F)$ were obtained at various heliocentric distances by radio-sounding. From these data, one can derive information on the plasma turbulence with the help of the theory of radio wave propagation in a statistically inhomogeneous medium. According to the theory, the spectrum of refraction coefficient fluctuations $G_n(q)$ is expressed as follows:

$$G_n(q) = C^2\left(q_0^2 + q^2\right)^{-p/2} \exp\left(-\frac{q^2}{q_m^2}\right) \tag{2.17}$$

Here p is the spectral index characterizing the turbulence regime of a medium, $q = 2\pi/l$ is the spatial wavenumber corresponding to inhomogeneities of the scale l, $q_0 = 2\pi/l_{max}$ is the wave number corresponding to the outer (maximum) spatial scale l_{max}, $q_m = 2\pi/l_{min}$ is the wave number corresponding to the inner spatial scale l_{min}. It is expected that the outer scale l_{max} is close to the solar radius. At the same time, the inner one l_{min} is small and characterizes the process of turbulence energy transformation into the heat. It is known that under both the fragmentation of turbulence vortexes in the terrestrial atmosphere and the random mixing of plasma streams or liquid ones, the Kolmogorov spectrum with $p \approx 11/3$ is often realized. The multiplier C characterizes the medium inhomogeneity intensity and is connected to the dispersion of refraction coefficient fluctuations σ_n^2. If the registration of medium inhomogeneities is experimentally possible for scales $l_{min} < l < l_{max}$, the spectrum of refraction coefficient fluctuations can be described as follows:

$$G_n = C^2 q^{-p} \tag{2.18}$$

Eq. (2.17) and Eq. (2.18) are phenomenological representations of plasma statistical inhomogeneities. Indeed, these do not describe the natural turbulence but rather the spectrum of various plasma statistical inhomogeneities such as inhomogeneities elongated along the magnetic field, small ejections of plasma, filamentary structures etc. Let us note that Eq. (2.18) is chosen by analogy with a neutral gas. Its application to the description of near-solar plasma inhomogeneities in the magnetic field has no theoretical substantiation. The technique of radio-sounding allows investigating plasma statistical inhomogeneities, which cause radio fluctuations. The representation of the spectrum of these fluctuations as Eq. (2.17) is phenomenological. Nevertheless, it allows determining four quantitative plasma characteristics: the spectral index characterizing the turbulence regime, the outer and the inner spatial scales of turbulence, and the relative plasma inhomogeneity. Indeed, using the term "turbulence" to describe statistical inhomogeneities is non-rigorous, but the magnetized plasma turbulence theory is still under development. Radio fluctuations are stipulated by the influence of plasma refraction coefficient statistical inhomogeneities δn connected to electron concentration fluctuations δN through the equation:

$$\delta n = -40.4 f^{-2} \delta N \qquad (2.19)$$

Here δN is measured in m^{-3} and f – in Hz. As is easy to see from Eq. (2.19), refraction coefficient spatial spectra (G_n) are identical to electron concentration ones (G_N). The authors (Armand et al. 1987, 135-141) (Armstrong and Woo 1981, 415-421) (Woo 1978, 727-739) (Woo et al. 1976a, 593-602) (Woo et al. 1976b, 568-574) (Bird 1982, 99-126) applied the theory of radio fluctuations in a statistically inhomogeneous medium to the analysis of radiofrequency and amplitude fluctuations in the solar wind. They accepted a few essential hypotheses. They supposed that plasma inhomogeneities situated on the ray path AB could be presented as the relatively thin "phase screen" D_1D_2 centered at the point C while parameters of the spectrum $G_n(q)$ depended only on the distance r_0 between the solar center O and the point C. Under such assumptions, the shaded region along the segment D_1D_2 (which delivers the main deposit to radio fluctuations) is situated at the heliocentric distance r_0 and the ‘‘phase screen” thickness is approximately equal to r_0 (Fig. 2-6).

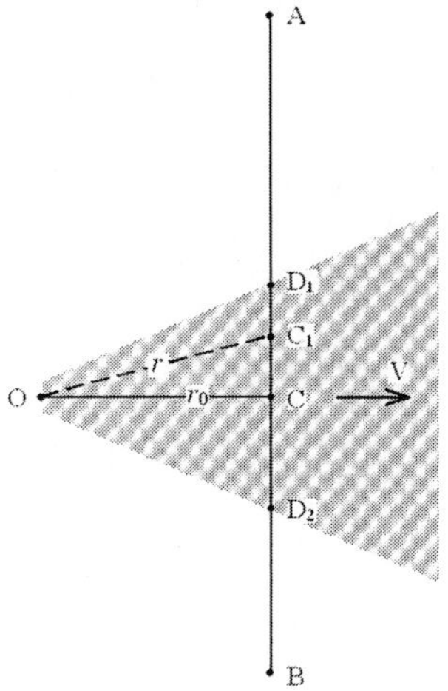

Fig. 2-6 Outline of plasma turbulence studies (Yakovlev and Pisanko 2018, 560)

The ‘‘phase screen” model has no distinctions between r_0 and r. Inhomogeneities were also considered as ‘‘frozen” into the solar wind (moving with the velocity V) and were assumed isotropic. In contrast, real inhomogeneities were radially elongated due to the magnetic field influence.

The ray path AB moves with the velocity $V_1 = V \pm dr_0/dt$, where V is the solar wind velocity, dr_0/dt is the velocity of approach of the ray path to the Sun (spacecraft’ recess behind the Sun – sign ‘‘plus”) or its motion away from the Sun (sign "minus"). Usually $V > dr_0/dt$ for $r > 8$, and one can suppose $V_1 \approx V$. The ray path intersects irregularities with the scale l during the time $\Delta t = l/V$; therefore, the fluctuation frequency F can be estimated as $F \approx V/l$. Low-frequency fluctuations are determined by large-scale plasma irregularities of the outer turbulence scale l_{max}. In contrast, high-frequency fluctuations depend on small-scale plasma irregularities with the scales exceeding the inner turbulence scale l_{min}. It was shown (Armand et al. 1987, 135-141) (Armstrong and Woo 1981, 415-421) (Woo 1978, 727-739) (Woo et al. 1976a, 593-602) (Woo et al. 1976b, 568-574) that radio amplitude fluctuations arose mainly from small-scale plasma inhomogeneities (with the scale of the order of the first Fresnel zone diameter $l_\varnothing$). So, the frequency spectrum of radio amplitude fluctuations $G_a(F)$ can be limited as follows:

$$V_1 / l_\varnothing \geq F \geq V_1 / l_{\min} \tag{2.20}$$

$$l_\varnothing = 2\left(\frac{\lambda l_1 l_2}{l_1 + l_2}\right)^{1/2}$$

Here l_1 = AC is the distance between the spacecraft and the point of the ray path closest to the Sun, l_2 = BC is the distance between the ground station and the point of the ray path closest to the Sun (Fig. 2-6). The theory of amplitude fluctuations is valid for weak fluctuations when the standard deviation $\sigma_a < 0.6$. Fluctuations of frequency f arise mainly from large scale plasma inhomogeneities, and its spectrum $G_f(F)$ has the maximum on low frequencies

$$V_1 / l_{\max} > F > V_1 / l_{\varnothing} \tag{2.21}$$

The theory of frequency fluctuations is valid for both weak and substantial fluctuations. In the case of experimental temporal spectra of both amplitude and frequency fluctuations obtained at the fulfillment of Eq. (2.20) and Eq. (2.21), it is possible to analyze these spectra with the help of Eq. (2.18), which describes the spatial spectrum of refraction coefficient irregularities $G_n(q)$. In this case, both the amplitude fluctuation spectrum and the frequency fluctuation spectrum can be presented by the power-law functions (Armand et al. 1987, 135-141) (Woo et al. 1976, 593-602) (Woo et al. 1976, 568-574) (Woo 1978, 727-739):

$$G_a = D_a F^{-\chi_a}, \chi_a = p - 1 \tag{2.22}$$

$$G_\varphi = D_\varphi F^{-\chi_\varphi}, \chi_\varphi = p - 1 \tag{2.23}$$

$$G_f = D_f F^{-\chi_f}, \chi_f = p - 3 \tag{2.24}$$

The spectral indexes of temporal spectra of amplitude fluctuations χ_a, phase fluctuations χ_φ, and frequency fluctuations χ_f depend on the plasma density turbulence spectral index p only. Multipliers D_a, D_φ, and D_f depend on the radio frequency f, the index of power p, and the heliocentric distance r. Exploring radio fluctuations in the near-solar plasma, one can obtain experimental spectra (2.22), (2.23), and (2.24), find the value of p, and determine dispersions of amplitude fluctuations σ_a^2, phase fluctuations σ_φ^2, and frequency fluctuations $\sigma_f^{\,2}$. The dispersions are determined by integration as follows:

$$\sigma^2 = \int_{F_{\min}}^{F_{\max}} DF^{-\chi} dF \qquad (2.25)$$

Here F_{max} and F_{min} are maximal and minimal frequencies of radio fluctuations for which experimental spectra $G_a(F)$, $G_\varphi(F)$, and $G_f(F)$ are available. To determine plasma turbulence characteristics, one should obtain experimental spectra $G_a(F)$, $G_\varphi(F)$, and $G_f(F)$ as well as dispersions σ_a^2, σ_φ^2, and $\sigma_f^{\,2}$ in the variation interval of F as wide as possible for various heliocentric distances.

The spatial spectrum of plasma inhomogeneities

The heliocentric distance dependence of the index of power p can indicate the turbulence regime change in the solar wind acceleration region. Let us note that the possibility of the spectrum presentation as Eq. (2.17) and Eq. (2.18) was not a priori known. The description of spectra $G_a(F)$ and $G_f(F)$ as power-law functions in the solar wind acceleration region was ascertained in radio-sounding experiments with the ''Mars 2" (1972) and the ''Mars 4" (1974) spacecraft. It was shown that Eq. (2.22) and Eq. (2.24) could describe the spectra, and the index of power p was estimated. The heliocentric distance dependence of the index of power p was revealed in radio-sounding experiments with the "Venera 10" spacecraft (1976). Based on spectra $G_a(F)$ and $G_f(F)$, it was shown that (within errors of measurement) the index of power p corresponded to the Kolmogorov spectrum ($p = 11/3$) for heliocentric distances $20 < r < 55$, and p decreased for $r < 6$. The estimates of the index of power p described in (Kolosov et al. 1982, 664-674) (Яковлев и др. 1977, 260-267) were derived from the data on the wavelength λ_1 of the deep space radio-communication system. Two research groups obtained estimates of p by analyzing both spectra of amplitude fluctuations (and phase fluctuations) and spectral line broadening in radio-sounding experiments with the ''Viking" spacecraft (1976) from the data on the wavelength λ_3 of the deep space radio-communication system. The detailed analysis of fluctuating signals and a significant number of radio transmission sessions allowed deriving the reliable heliocentric distance dependence of p (Armstrong and Woo 1981, 415-421) (Tyler et al. 1981, 318-332). The heliocentric distance dependence of the index of the power of the plasma inhomogeneity spatial spectrum is shown in Fig. 2-7A (Armstrong and Woo 1981, 415-421).

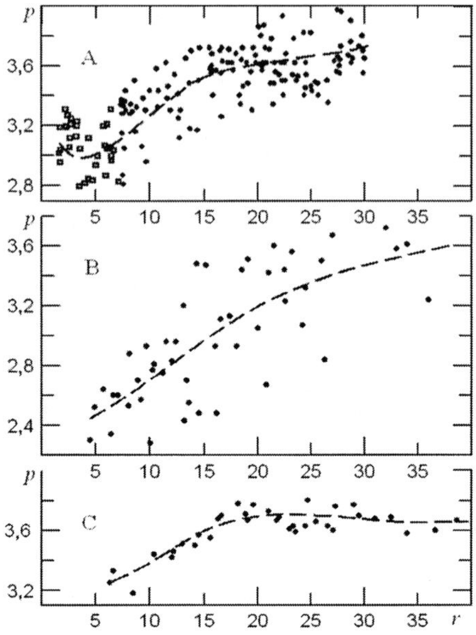

Fig. 2-7 Plasma inhomogeneity spectral index (p) heliocentric distance dependencies; the heliocentric distance dependence of p derived by Armstrong and Woo (1981, 415-421) is shown in A; the heliocentric distance dependence of p derived from amplitude fluctuation spectra for heliocentric distances $3 < r < 60$ obtained from ‘‘Venera 15” and ‘‘Venera 16” data is shown in B; the heliocentric distance dependence of p derived from spectra of radiofrequency fluctuations at the wavelength λ_3 obtained from ‘‘Galileo” data is shown in C (Yakovlev and Pisanko 2018, 561)

The ‘‘Venera 15” and the ‘‘Venera 16” spacecraft emitted radio waves at the wavelengths λ_1 and λ_2. This allowed deriving the heliocentric distance dependence of p from amplitude fluctuation spectra for heliocentric distances $3 < r < 60$ (Ефимов и др. 1987, 2025-2032) (Яковлев и др. 1988, 1290-1299). The dependence is shown in Fig. 2-7B. Radio-sounding with the ‘‘Galileo” (1999) spacecraft allowed deriving the heliocentric distance dependence of p from the analysis of spectra of radiofrequency fluctuations at the wavelength λ_3 (Efimov et al. 2005, 1448-1453). The dependence is shown in Fig. 2-7C. Similar heliocentric distance dependencies were obtained in radio-sounding experiments with subsequent spacecraft (Chashei et al. 2005, 1454-1460) (Ефимов и др. 2010, 1343-1349).

Studies of spectra of radio fluctuations showed that the value of p corresponds to the Kolmogorov spectrum (p = 3.66) for $r \geq 25$ and the average value of $p = 3.0 \pm 0.2$ for $3 < r < 5$. Fig. 2-7 testifies for the plasma turbulence regime change at heliocentric distances $5 < r < 18$. Here the fundamental acceleration of the solar wind takes place, and the stream structure can already be formed (Pisanko 1996, (3)61-(3)64). The beam instability created here due to the interaction between solar wind streams with different velocities changes the turbulence regime.

One should obtain experimental spectra of radio fluctuations for ultra-low frequencies F to determine the outer scale l_{max} of plasma inhomogeneities. It is possible in the registration of high stable signals during a long time interval: only the registration of frequency fluctuations during long time intervals allows to obtain reliable low-frequency spectra $G_f(F)$. In experiments with the "Galileo" (1995) and the "Ulysses" (1991) spacecraft, the continuous registration of frequency fluctuations during more than 10 hours took place, and this allowed to obtain spectra $G_f(F)$

reliable in the frequency range from $F \approx 10^{-1}$ Hz to $F \approx 3\times10^{-5}$ Hz (Chashei et al. 2007, 611-620) (Efimov et al. 2002, 579-590) (Ефимов и др. 2010, 492-502) (Рамзанов и др. 1979, 1051-1060). The peculiarities of the spectra are the maximum of $G_f(F)$ at $F \approx 10^{-4}$ Hz and the decrease of $G_f(F)$ at lower fluctuation frequencies. According to Eq. (2.17), the maximum of $G_f(F)$ exists at low fluctuation frequency and corresponds to the frequency (Efimov et al. 2002, 492-502) (Ефимов и др. 2002, 640-650):

$$F_{\min} = V l_{\max}^{-1} \left(\frac{2}{p-3} \right)^{1/2} \qquad (2.26)$$

Efimov, with colleagues (Chashei et al. 2007, 611-620) (Ефимов и др. 2010b, 492-502), note that one can estimate (with the help of Eq. (2.26)) the outer scale l_{max} from known V, p, and measured F_{min}. The heliocentric distance dependence of l_{max} is shown in Fig. 2-8; points correspond to the ‘‘Galileo” spacecraft (1995) signals, squares – to the ‘‘Ulysses” (1991) ones.

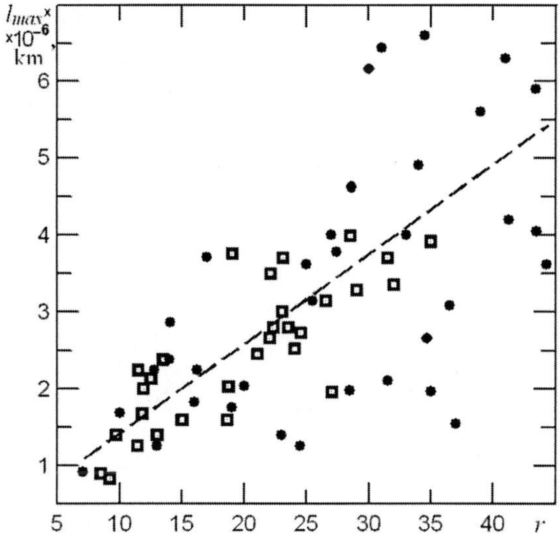

Fig. 2-8 Outer scale of plasma turbulence l_{max} at various heliocentric distances; filled circles correspond to data obtained with the ''Galileo'' spacecraft (1995); open squares – to the ''Ulysses'' (1991) (Yakovlev and Pisanko 2018, 562)

Despite the wide spacing of l_{max}, the averaged heliocentric distance dependence of l_{max} can be indicated as follows:

$$l_{\max} \approx 1.2 \cdot 10^5 r[km]$$

It is shown by the dashed line in Fig. 2-8; $l_{max} = 1.5\times10^6$ km for $r = 10$, $l_{max} = 3\times10^6$ km for $r = 25$. The solar corona large-scale spatial structure determines the outer scale of turbulence of the order of the solar diameter.

One should explore experimental spectra of radio fluctuations on sufficiently high frequencies F to study small-scale plasma inhomogeneities and, in particular, determine the inner scale l_{min}. Exploited were experimental data on both spectra of amplitude fluctuations (and frequency fluctuations) and the spectral line broadening when fluctuating signals on frequencies $F > 10$ Hz were registered. First, the spectral line λ_1 broadening from the "Venera 10" spacecraft (1976) was exploited (Рамзанов и др. 1979, 1051-1060). The authors showed that the spectral line shape contained new information on small-scale irregularities under large deviations δf from the line central frequency. Using the maximum frequency deviation (F_{max}) from the line central frequency, it is possible to estimate the minimum scale of irregularities $l_{min} = V_1 F_{max}^{-1}$; the heliocentric distance dependence of l_{min} is depicted by squares in Fig. 2-9.

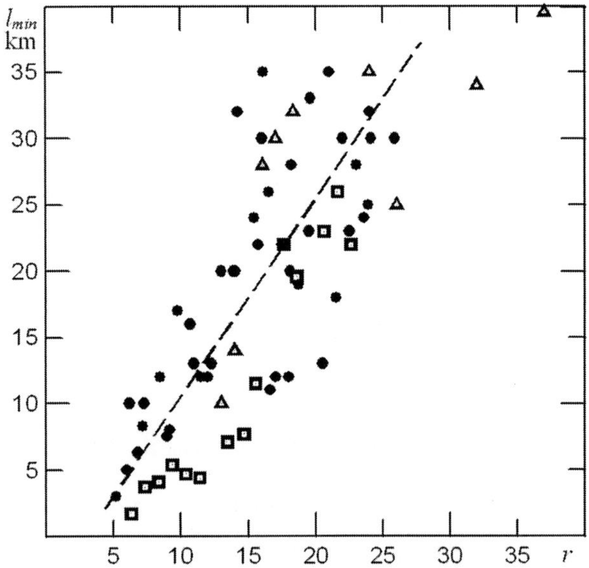

Fig. 2-9 Plasma turbulence inner scale l_{min} at various heliocentric distances; the values of l_{min} are shown as open squares (Рамзанов и др. 1979, 1051-1060), open triangles (Ефимов и др. 2010b, 492-502), and filled circles (Яковлев и др. 1988b, 1290-1299); the averaged heliocentric distance dependence is shown as the dashed line (Yakovlev and Pisanko 2018, 562)

Frequency fluctuations at high frequencies F were measured in radio-sounding experiments with the "Nozomi" (2000) spacecraft, and F_{max} along with reliable spectra $G_f(F)$ were obtained (Ефимов и др 2010, 492-502). The authors used the solar wind velocity V measured in experiments with other spacecraft to determine G_{min}. The obtained values of l_{min} are depicted by triangles in Fig. 2-9.

Another way to determine l_{min} was the analysis of amplitude fluctuations in radio-sounding experiments with the ‘‘Venera 15” and the ‘‘Venera 16” (1984) spacecraft (Яковлев и др. 1988, 1290-1299). Registered were amplitude fluctuations at wavelengths λ_1 and λ_2. Under the arbitrary assumption that the spectral intensity G_a of for sure registered maximal frequency F_{max} is a hundred times smaller than the spectral intensity registered at low frequency, the authors find l_{min} from known V. The obtained values of l_{min} are depicted by points in Fig. 2-9. It is clear from Fig. 2-9 that various methods give close values of the turbulence inner scale l_{min}: l_{min} = 13 km for r = 10, l_{min} = 26 km for r = 20. The averaged heliocentric distance dependence $l_{min} \approx 1.3\ r$ [km] is shown by the dashed line in Fig. 2-9. The inner turbulence scale is of the order of the proton gyroradius. It seems that the proton gyroradius is the minimal inhomogeneity scale where the turbulence energy transfers into the solar wind additional heating.

Radial variations of radio fluctuation intensity and fractional density fluctuations

It is desirable to ascertain the plasma relative inhomogeneity heliocentric distance dependence to understand the solar wind strong acceleration. Let us characterize the radio fluctuation intensity by standard deviations of the amplitude σ_a and the frequency σ_f, while the relative plasma inhomogeneity (the amplitude of the fractional density fluctuation) – by the ratio of electron concentration standard deviation σ_N to electron concentration value $\gamma = \sigma_N/N$. Parameters σ_a, σ_f, σ_N are connected and depend on the heliocentric distance.

Let us analyze experimental heliocentric distance dependencies of σ_a and σ_f. During radio-sounding, the frequency is always registered every second and only in selected space missions the amplitude was registered (15–30) times per second. σ_a and σ_f are calculated from these data. Various experimental heliocentric distance dependencies of σ_a are published in (Ефимов и др. 1987, 2025-2032) (Ефимов и др. 2008, 1257-1268) (Колосов и др. 1978, 555-558) (Рубцов и др. 1987, 620-625) (Яковлев и др. 1988, 3-9) (Яковлев и др. 1988, 1290-1299), while heliocentric distance dependencies of σ_f – in (Armand et al. 1987, 135-141) (Efimov et al. 1977, 172-180) (Efimov et al. 2002, 453-458) (Efimov et al. 2008, 117-123) (Efimov et al. 2014, 1715-1729) (Ефимов и др. 1977, 322-325)

(Ефимов и др. 2010, 1343-1349) (Ефимов и др. 2013, 17-27) (Ефимов и др. 2013, 916-925) (Колосов и др. 1978, 555-558). The authors of listed papers use power-law functions to fit experimental dependencies σ_a and σ_f:

$$\sigma_a \propto r^{-\alpha}$$

$$\sigma_f \propto r^{-\beta} \tag{2.27}$$

and determine the values of parameters α and β. The results are presented in Tables 2-1 and 2-2. The first line contains the spacecraft and time, the second line – the wavelength, the third line – the heliocentric distance interval, the fourth line – values of α and β.

Table 2-1 Values of α derived from amplitude fluctuations

spacecraft	Helios, 1975	Voyager, 1979	Voyager, 1997	Voyager, 1980	Voyager, 1980	Venera, 1984
λ cm	13	3,6	13	13	3,6	5
r	11÷30	6÷17	12÷23	12÷30	12÷30	5÷27
α	1,55	1,82	1,87	1,73	1,89	1,73

Table 2-2 Values of β derived from carrier frequency fluctuations

spacecraft	Helios, 1975	Pioneer, 1975	Viking, 1976	Pioneer, 1981	Venera, 1984	Nozomi, 2000
λ cm	13	13	13	13	32; 5	13
r	3÷50	7÷60	3÷27	15÷60	4÷40	5÷35
β	1,79	1,59	1,87	1,66	1,65	1,61

The average α = 1.75 ± 0.15 for heliocentric distances 6 < r < 30 was obtained from the amplitude data while the average β = 1.70 ± 0.17 for heliocentric distances 5 < r < 40 – from frequency data. The listed above publications presented experimental heliocentric distance dependencies of σ_a and σ_f that were obtained in six cycles of coronal radio-sounding experiments.

The determination of the heliocentric distance dependence of the fractional density fluctuation amplitude $\gamma(r)$ needs the electron concentration standard deviation heliocentric distance dependence $\sigma_N(r)$, which can be found from amplitude fluctuations or frequency ones. As shown theoretically in (Armand et al. 1987, 135-141) (Armstrong and Woo 1981, 415-421) (Efimov et al. 1977, 172-180) (Woo 1978, 727-739) (Woo et al. 1976, 593-602) (Woo et al. 1976b, 568-574), the determination of $\sigma_N(r)$ from $\sigma_a(r)$ faces difficulties. $\sigma_f(r)$ is more informative: it works for both intense and weak fluctuations and covers the longer heliocentric distance interval. One can determine $\sigma_f(r)$ both by integration of frequency fluctuation spectra (Eq. (2.24) and Eq. (2.25)) and from the width of spectral line broadening $\Delta f \approx 2.5\sigma_f$.

Theoretical approximations of the relationship between σ_f and σ_N are published in (Armand et al. 1987, 135-141) (Kolosov et al. 1982, 664-674)

(Woo 1978, 727-739). The more convenient expression is presented in (Efimov et al. 2008, 117-123) (Ефимов и др. 2010, 1343-1349):

$$\sigma_f \propto \sigma_N \left[\left(\frac{\chi_f}{1-\chi_f} \right)^{1/2} L_e^{1/2} l_{\max}^{-\chi_f/2} V^{(\chi_f+1)/2} \right] \quad (2.28)$$

Here $L_e \approx r_0 \approx D_1D_2$ is the conventional phase screen thickness, χ_f is the frequency fluctuation spectrum index, V is the velocity of the plasma inhomogeneity transfer through the ray path, l_{max} is the outer scale of inhomogeneities. If we do not take into account cumbersome multipliers, which have no heliocentric distance dependence, put $L_e \approx r_0$, $\chi_f = p$–3, exploit the experimental heliocentric distance dependence $l_{max}(r)$ and Eq. (2.27) for σ_f, then (according to Eq. (2.28)) we obtain:

$$\sigma_N \propto r^{-\beta} \left[\left(\frac{p-3}{4-p} \right)^{-1/2} r^{(p-4)/2} V^{(2-p)/2} \right] \quad (2.29)$$

According to Fig. 2-7, the plasma spatial inhomogeneity spectrum index p weakly depends on the heliocentric distance for $15 < r < 40$. Therefore, let us put $p \approx 3.5$. One obtains (from Eq. (2.29) for the average value $\beta = 1.7$ (Table 2-2)):

$$\sigma_N \propto r^{-1.95} V^{-1.75} \quad (2.30)$$

Eq. (2.30) is valid when σ_f is determined from very long records of frequency fluctuations. If σ_f is determined from a twenty-minute record of frequency fluctuations, then the heliocentric distance dependence of l_{max} does not manifest itself, and one should put $l_{max} = const$ that leads (from Eq. (2.28)) to:

$$\sigma_N \propto r^{-2.2} V^{-1.75} \tag{2.31}$$

According to Eq. (2.5), Eq. (2.6), Eq. (2.12), and Eq. (2.13), $V = const$ and $N \infty r^{-2}$ for $r > 20$; hence, the fractional density fluctuation amplitude γ will be almost constant for $15 < r < 30$ owing to Eq. (2.30) and Eq. (2.31). Eq. (2.30) and Eq. (2.31) imply that the mean frequency deviation has a slightly different radial falloff, depending on the specific band in the fluctuation spectrum. The longer period fluctuations decrease more rapidly with r than the shorter period fluctuations. For determining the fractional density fluctuation amplitude γ in the solar wind acceleration region, numerical calculations are needed to take into account heliocentric distance dependencies $V(r)$, $N(r)$, $p(r)$. Based on Eq. (2.5), Eq. (2.6), Eq. (2.12), and Eq. (2.13) under $\beta = 1.7$, the calculations showed that the fractional density fluctuation amplitude γ increased with the heliocentric distance in the solar wind acceleration region. Experimental values of β varied from 1.5 to 1.9 for various space missions; therefore, various heliocentric distance dependencies of the fractional density fluctuation amplitude $\gamma(r)$ (from almost independent on the heliocentric distance till the increase with the heliocentric distance) could be realized. Estimates of amplitudes of fractional density fluctuation heliocentric distance dependencies derived from heliocentric distance dependencies of frequency fluctuations $\sigma_f(r)$ and amplitude ones $\sigma_a(r)$ under various assumptions about solar wind density heliocentric distance dependencies $N(r)$ and solar wind velocity ones $V(r)$ in different years are collected in (Chashei et al. 2007, 611-620) (Efimov et al. 1977, 172-180) (Imamura et al. 2014, 117-135) (Рубцов и др. 1987, 620-625) (Woo et al. 1995, 329-332) (Яковлев и др. 1988, 1290-1299). Let us note that heliocentric distance dependencies of the fractional density fluctuation amplitude are doubtful for both $r < 5$ (where radio fluctuations are large) and $r > 40$ (where the fluctuations are small). According to the publications, values of γ variate strongly both at small heliocentric distances and large ones. The fractional density fluctuation amplitude dispersion remains acceptable inside the heliocentric distance interval $6 < r < 40$. Heliocentric distance dependencies of the fractional density fluctuation amplitude obtained from radio-sounding experiments (Imamura et al. 2014, 117-135) (Рубцов и др. 1987, 620-625) (Яковлев и др. 1988, 1290-1299) with the ‘‘Venera 15” and the ‘‘Venera 16” spacecraft (1984) and with the ‘‘Akatsuki” spacecraft (2011) are shown in Fig. 2.10.

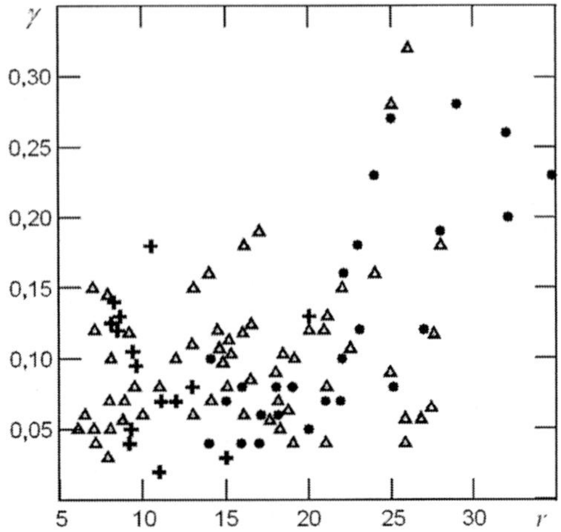

Fig. 2-10 Low values of fractional density fluctuations for $5 < r < 20$; filled circles and open triangles correspond to the frequency and amplitude data from the "Venera 15" and the ''Venera 16"; crosses correspond to frequency data from the ''Akatsuki". (Yakovlev and Pisanko 2018, 564)

Points and triangles in Fig. 2-10 correspond to the frequency and amplitude data from the "Venera 15" and the ''Venera 16" spacecraft while crosses – to frequency data from the ''Akatsuki" spacecraft. The fractional density fluctuation amplitude $\gamma(r)$ increase with the heliocentric distance is evident: $\gamma = 0.08$ for $r = (15–6)$ while $\gamma = 0.2$ for $r = (30–20)$. Fractional density fluctuation amplitude values obtained from radio-sounding experiments (Chashei et al. 2007, 611-620) (Efimov et al. 1977, 172-180) with the ''Venera 10" spacecraft (1978) and the ''Galileo" spacecraft (1997) are shown in Fig. 2-11.

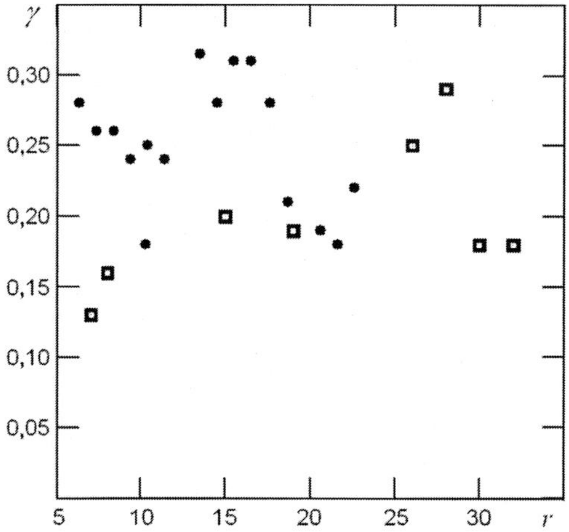

Fig. 2-11 High values of fractional density fluctuations in the solar wind acceleration region; filled circles correspond to data from the ‘‘Venera 10” spacecraft; open squares - to the ‘‘Galileo” spacecraft (Yakovlev and Pisanko 2018, 564)

One can see the heliocentric distance independence (within the errors of measurements) of the fractional density fluctuation amplitude: its average value is equal to $\gamma = 0.25$. Woo, with colleagues (Woo et al. 1995, 329-332), concluded that the fractional density fluctuation amplitude $\gamma(r)$ was (a) essentially independent on heliocentric distance in the slow solar wind, and (b) factor thirty smaller in the fast wind than in the slow wind but increased dramatically with the heliocentric distance.

The search of wavelike structures

Quasi-periodic carrier frequency variations should be registered when radio-sounding plasma density waves. The effect will be masked by random carrier fluctuations stipulated by statistical plasma inhomogeneities. Efimov, with colleagues (Ефимов и др. 1990, 581-587) (Ефимов и др. 1993, 1075-1085) (Ефимов и др. 1993, 143-152) (Efimov et al. 1994, 493-496), attempted to derive quasi-periodic signals in the near-solar plasma from the data of "Venera 15-16" (1984), "Viking" (1976), and "Voyager" (1979) missions by analyzing spectra of both carrier frequency fluctuations and amplitude ones. They draw attention to the peculiarity of the velocity heliocentric distance dependence determined by Eq. (2.1) from carrier amplitude fluctuations and consider the solar wind velocity seeming sharp decrease at $r \approx 12–16$ as the manifestation of plasma density waves in this heliocentric distance interval. Based on these ideas, Yakubov, with colleagues (Якубов и др. 1989, 772-775), studied peculiarities of frequency (λ_1) variations at heliocentric distances between 6 and 30. For the confident detection of quasi-periodic frequency variations, they exploited simultaneous measurements at two ground receiving stations. Used were 20 to 30-minute communication sessions of second timing frequency measurements each. Quasi-periodic frequency variations were revealed, but the effect was strongly variable from a session to a session. They interpreted the detected quasi-periodic frequency variations as the manifestations of quasi-periodic electron concentration variations on the ray path. The latter could be plasma density waves arising due to the beam instability in the interaction region of solar wind streams with different speeds.

Efimov, with colleagues (Ефимов и др. 2011, 1433-1443) (Efimov et al. 2012, 500-5078) (Ефимов и др. 2013, 916-925), used the data on the reduced frequency difference $\Delta f_{3,4}$ of two coherent radio waves λ_3 and λ_4. This difference is proportional to the time derivative of the electron content on the ray path and does not depend on the carrier frequency regular Doppler shift. It allows more accurate inspection of its quasi-periodic variations. They exploited the data from "Ulysses" (1995), "Mars-Express" (2006), "Venus-Express" (2006), and "Rosetta" (2006) missions registered at four ground receiving stations. Three processing techniques were applied to detect the periods: autocorrelation analysis and spectral and wavelet ones. The heliocentric distance interval varied from 3 to 38. The thorough search revealed quasi-periodic frequency variations in 20% of cases; the period varies randomly from 2 minutes to 4 minutes. The phenomenon was essentially nonstationary: it was observed for several tens of minutes and then disappeared. Detected periods corresponded to known 5-minute oscillations detected optically in the low corona. They supposed that detected were magneto-sound waves created by Alfven waves propagating from the low corona outside. Additional information was obtained from the "Akatsuki" spacecraft (Miyamoto et al.

2014, 51-58). Periods of plasma density waves (in seconds) are gathered in Fig. 2-12.

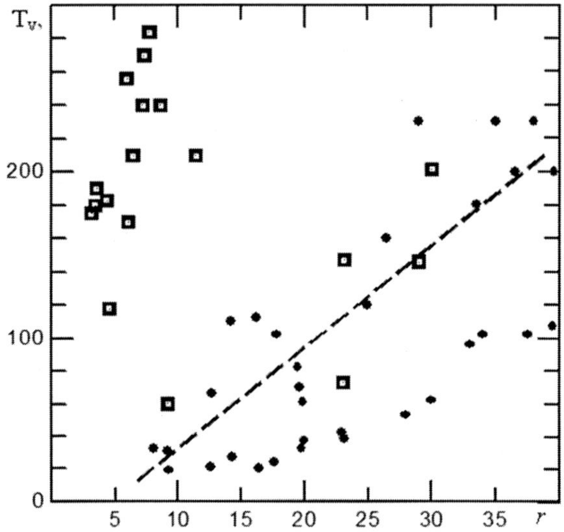

Fig. 2-12 Periods of plasma density waves at various heliocentric distances

Data of (Ефимов и др. 2011, 1433-1443) (Efimov et al. 2012, 500-508) are depicted by squares, data of (Ефимов и др. 2013, 916-925) – by points. The period T at λ_1=32 cm increases from 20 seconds to 140 seconds for the heliocentric distance interval r = 6–30, while at λ_3=13 cm – $T \approx$ 180–300 seconds for the heliocentric distance interval r = 3–5. One can conclude that there are quasi-periodic variations of two types. The first type ($T \approx$ 180–300 seconds) has no evident dependence on the heliocentric distance and probably is excited by 5-minute photosphere oscillations. The second one has the explicit dependence on the heliocentric distance and is excited locally due to the interaction of solar wind streams with different speeds.

Alfven waves can be detected with the help of Faraday's effect when linearly polarized radio waves are used. "Helios 1-2" (1975) spacecraft provided the radio sounding with linearly polarized radio waves λ_3. The variations of the Faraday polarization angle Φ were registered at several ground stations. The equation describes Faraday's effect:

$$\Phi(r_0,t) = Kf^{-2}\int_A^B N(r,t)H_l(r,t)dl \qquad (2.32)$$

Here $K = 1.35\times10^{-8}$, Φ is expressed in degrees, f is expressed in MHz, N is expressed in cm^{-3}, dl is the element on the ray line AB (in cm) of integration, H_l is the magnetic field stress projection on the ray line AB (in Gauss), r is the heliocentric distance of an arbitrary point C_1 of the ray line, r_0 = OC is the minimal distance to the ray line AB (Fig.2-6). For the spherically symmetric electron concentration $N(r,t)$ and the magnetic field $H(r,t)$, Φ = 0 because integrals along the ray line parts CA and CB are equal but have opposite signs. In the registered data, the almost regular variation of the averaged dependence $\Phi(r_0)$ was revealed on which fast fluctuations $\delta\Phi(t)$ were superimposed. Efimov, with colleagues (Ефимов и др. 1993, 1075-1085) (Ефимов и др. 1993, 916-925), assumed that the analyses of the Faraday angle fluctuations could indicate the presence of Alfven waves. They analyzed experimental dependencies $\Phi(t)$ obtained in two communication sessions using the polarized signal of the "Helios 2" (1976) spacecraft when the parameter r_0 was equal to 2.8 and 3.4. They eliminated the slowly varying component $\Phi(t)$ by moving mean. The experimental realization of the random function $\delta\Phi(t)$ with zero average value and sampling rate of 1 Hz was constructed. The correlation function and the corresponding spectrum were calculated to extract the quasi-periodic component. Almost periodic variations of the correlation function were found, and the characteristic period of variations was determined. To enhance the reliability of these results, the variations $\delta\Phi(t)$ registered simultaneously at two ground receiving stations were analyzed. The analysis showed the same effects at two spaced ground receiving stations. Slow quasi-periodic variations of the Faraday angle are evident under averaging $\delta\Phi(t)$ over twenty-second-time intervals. Efimov, with colleagues (Ефимов и др. 2013, 17-27), demonstrated examples of such variations: first, variations with the period of 2.6 minutes were observed during 10 minutes on the rising background, and then variations with the period of 3.9 minutes were observed during 20 minutes. The amplitude of variations achieved $\pm 2^{\circ}$ that testified on the irregular presence of Alfven waves. When interpreting these observations, the need to compare the deposits of electron concentration (δN) variations and magnetic field (δH) ones arose. From Eq.(2.32), one can write:

$$\delta\Phi(r_0,t) = Kf^{-2}\int_A^B \delta N(r,t)H_l(r,t)dl +$$

$$Kf^{-2}\int_A^B N(r,t)\delta H_l(r,t)dl$$

Here, the first addendum describes the influence of electron concentration on Faraday's effect, and the second – the influence of magnetic field variation. The authors assigned the revealed quasi-periodic variations to fast magnetic variations (described by the second addendum) at the ray path. They concluded about the influence of Alfven waves on the linearly polarized radio waves. Efimov, with colleagues (Efimov et al. 2015, 2397-2408) (Ефимов и др. 1999, 312-320), presented the additional information on random and periodic variations of the Faraday angle and calculated its spectrum $G_\Psi(F)$ as the power-law dependence: $G_\Psi(F) \propto F^{-\chi_\Psi}$ in the frequency range 3MHz > F > 1.7 MHz. Here χ_Φ was the spectral index of fluctuations Φ which they found to be 1.9 for r_0 = 3–4 and 1.2 for r_0 = 6–10. The Faraday angle variation periods were 80–300 seconds. The analysis of fluctuations $\delta\Phi(t)$ registered simultaneously at two spaced ground stations allows us to estimate the velocity (V). Significant velocity variations (from 100 km s^{-1} to 1000 km s^{-1}) were found at $r \approx 5.6$ during an hour. The authors supposed that registered velocities $V = 600$–1000 km s^{-1} corresponded to the Alfven wave speed, not to the solar wind bulk speed. Jensen, with colleagues (Jensen et al. 2013, 71-82), analyzed the Faraday angle variations obtained during the “Messenger” (2009) mission and found quasi-periodic variations (with periods of 4.2 minutes and 14 minutes at $r \approx 3.2$) in addition to periodic and random ones.

Magnetic field estimates from radio sounding with spacecraft signals

Several authors (Levy et al. 1999, 596-600) (Stelzried et al. 1970, 440-456) (Volland et al. 1977, 659-672) (Patzold et al. 1987, 91-105) attempted to explore the magnetic field at heliocentric distances between 3 and 14 with the help of Faraday’s effect (Eq. (2.42)). The effect is evident in the case of no spherical symmetry of whether the electron concentration or the magnetic field. The first explorations of the Faraday angle on the data from “Pioneer 6” (1968) and “Pioneer 9” (1970) missions brought an unexpected result: for $r_0 = 6$–7, the angle demonstrated the significant slow variation of 20°–30° during two hours at the carrier frequency 2.29 GHz that reflected the magnetic field sector structure in the solar wind acceleration region (Levy et al. 1999, 596-600) (Stelzried et al. 1970, 440-456). Bird, with colleagues (Volland et al. 1977, 659-672) (Patzold et al. 1987, 91-105), found significant regular variations of the Faraday angle in the data from “Helios 1-2” spacecraft and derived the dependence $\Phi(r_0)$. They assumed that the dependence was the manifestation of magnetic sectors.

Ness and Wilcox discovered the interplanetary magnetic field sector structure at the Earth orbit in 1965 (Ness and Wilcox 1965, 1592-1594). It is the typical behavior of the direction of averaged over three-hour intervals interplanetary magnetic field component in the ecliptic. During approximately a week (or two), the component is directed toward the Sun (away from the Sun) and then the direction quickly (during a few minutes) changes to the opposite. The structure recurs within ~ 27 days (the period of rotation of solar equatorial regions).

The electron concentration heliocentric distance dependence was supposed as follows:

$$N = N_0 r^{-\xi}$$

Parameters N_0 and ς were known. The magnetic field stress heliocentric distance dependence was supposed as follows:

$$H = H_0 r^{-\xi_h}$$

Parameters H_0 and ξ_h were unknown. To best fit the experimentally measured curve for the Faraday angle by varying the number of magnetic sectors and unknown parameters H_0 and ξ_h, they obtained the empirical magnetic field heliocentric distance dependence as follows:

$$H = 7.9r^{-2.7} \tag{2.33}$$

or

$$H = 6r^{-3} + 1.2r^{-2} \tag{2.34}$$

H is expressed in Gausses in Eq. (2.33) and Eq. (2.34). Both equations are valid for $r \approx 4$–11. The complexity of the applied procedure does not allow estimating its errors.

Summary of Chapter 2

An overview of coronal radio-sounding data is presented, applicable to heliocentric distances $3 < r < 60$, heliolatitudes below 60°, and minimum to moderate solar activity. The heliocentric distance dependence of the solar wind velocity is presented. The accuracy of the mean radial dependence, based on data from four space missions, is estimated to be less than 20%. There is insufficient data at heliocentric distances below 4, where the error increases to about 40%. Eq. (2.5) and Eq. (2.6) are satisfactory approximations to the solar wind velocity heliocentric distance dependence for $r > 3$. The main acceleration of the solar wind takes place at heliocentric distances $7 < r < 24$, over which the solar wind velocity increases by a factor of seven. The electron density decreases as r^{-2} at heliocentric distances $r > 24$ but falls off more rapidly closer to the Sun. A satisfactory approximation for the electron density heliocentric distance dependence at $r > 3$ is given by Eq. (2.12) and Eq. (2.13), assuming the solar wind flux conservation. From Eq. (2.5), Eq. (2.6), Eq. (2.12), and Eq. (2.13), it may be concluded that the solar wind power increases by a factor of 30 over the radial distance from 6 to 30; the maximum acceleration of 2.6 m s^{-2} achieves at the heliocentric distance $r = 16 \pm 2$. A dramatic change in the solar wind turbulence regime occurs at heliocentric distances in the interval $8 < r < 20$. The spectral power index of electron density fluctuations is equal to 3.66 ± 0.1 at distances $r > 20$ (corresponding to the Kolmogorov spectrum), but decreases to 3 ± 0.2 at heliocentric distances $3 < r < 5$. The inner scale of density inhomogeneities is proportional to the heliocentric distance, increasing from 10 km at $r = 10$ up to 32 km at $r = 25$. This order of magnitude corresponds to the proton gyroradius. The continuous registration of radio wave frequency fluctuations over a long time allows determining the outer scale of density inhomogeneities. It is also directly proportional to the heliocentric distance, reaching a value of 1.5×10^6 km at the heliocentric distance $r = 10$. Estimates of the amplitude of fractional density fluctuations ($\gamma = \sigma_N/N$) derived from coronal radio-sounding observations are approximations because the electron density (N) and its standard deviation (σ_N) were determined during different periods.

The amplitude of the fractional density fluctuation can vary from the minimum ~ 0.05 to the maximum ~ 0.25. The heliocentric distance dependence in fast solar wind may differ from the slow solar wind one.

CHAPTER 3

SOLAR WIND: DIRECT MEASUREMENTS ON SPACECRAFT

We describe the results of solar wind plasma measurements on spacecraft with the help of modulation traps, electrostatic analyzers, and magnetometers. Principles of operation of the listed devices are explained without the association to specific models. The procedure of determining the solar wind plasma parameters from in-situ measurements of electric currents in the interplanetary medium exploited for data processing of the "Prognoz 7" spacecraft is described in detail as an example.

Collected are measured velocity distribution functions of solar wind protons, electrons, and α-particles. Often these differ from Maxwell ones. Measured parameters of the solar wind turbulence are reviewed.

Statistics of local solar wind plasma and interplanetary magnetic field measurements along trajectories of numerous spacecraft gathered during more than 50 years is analyzed.

Considered are discontinuities and their "images" in the data of in-situ measurements of the solar wind plasma. The traditional classification of discontinuity types is presented as a reference.

Devices for direct measurements of electric currents and magnetic fields in the solar wind plasma

One reestablishes solar wind plasma parameters from in-situ measurements of electric currents in the solar wind along a spacecraft trajectory. The flux and the energy spectrum of solar wind components in the interval of energies between 100 eV and 20 keV, creating cosmic electric currents, are calculated from plasma probe measurements. Usually, the Faraday cylinders (ion traps) and electrostatic analyzers are used as plasma probes. As the Debye radius (which is of the order of 10 m in the solar wind plasma at the Earth orbit) is much more than a spatial probe scale, the separation between electrons and protons can be realized with the help of electric fields. Detected electric currents in the solar wind are weak because of the low plasma concentration.

The typical Faraday cylinder consists of the input net (grounded to the spacecraft body) and the collector. One applies positive or negative voltages to the collector to separate coming electrons or ions. The collector current determines the flux of corresponding particles because $I=\mathrm{e}nVA$, where A is the collector area, nV is the flux of particles coming to the collector with energies

$$\frac{1}{2}m_iV^2 > ZeU$$

Here m_i is the ion mass, Z is the ion charge, U is the bias voltage. For the plane detector, the effective electric current on the collector depends on the entering particle direction; for angles of arrival $\theta < 45^\circ$, the current is approximately $I_n\cos\theta$ where I_n is the current in case of the direction of coming particles normal to the collector. For angles smaller than 30°, the error due to unknown (or variable) angle of incidence is ~ 15%. Spherical and hemispherical probes impair the energy resolution of measurements in the solar wind, whose bulk velocity is much more than ion thermal velocities.

There is an electric potential difference between the spacecraft body and the solar wind plasma, which changes the effective decelerating voltage for solar wind particles coming to the device. The difference is initiated either by the photoelectron emission from the spacecraft body (in case of the positive electric charge of the spacecraft) or high chaotic velocities of electrons compared to ions (in case of the negative electric charge of the spacecraft). The relative importance of the effects is unknown, and this impedes estimating the electric potential difference between the spacecraft and the solar wind. It is assumed that this variable difference is of the order of several volts (the bulk motion energy of solar wind electrons). However, because the electron thermal speed is always much more than its bulk speed, unknown varying electric potential differences complicate determining the bulk speed based on electron energy spectra measurements. Also, in most cases, the collector electric current is unsaturated, which impedes precise solar wind electron measurements. Usually, one measures the ion electric current with the help of the Faraday cylinder and restores the bulk speed, the concentration, and the temperature of the solar wind ions. In this case, the electric potential difference can be neglected because it is much less than the ion bulk motion energy.

The main uncertainty of ion electric current measurements is connected with photoelectrons knocked out from the collector by solar photons and plasma electrons and ions (the secondary electron emission). Suppose the device looks to the Sun (the usual mode of operation). In that case, the photoelectric current can be significant: the estimate of the solar photon flux out of the terrestrial atmosphere leads to the photoelectron emission from any metallic surface of the order of 10^{-8} A cm^{-2}. Such a photoelectric current stipulates the relatively low sensitivity of the ion flux measurements (the low limit ~ 10^{11} cm^{-2} s^{-1}). Photoelectric currents can be one hundred times diminished when an additional net is mounted between the input net and the collector. A negative bias voltage is applied to the additional net relative to the input net and the collector. This intermediate net impedes leaving photoelectrons from the collector. However, solar photons can be reflected from the collector and initiate the photoelectron emission from the intermediate net backside. Such photoelectrons can reach the collector, decreasing the ion current. The smaller is the intermediate net surface to collector surface ratio, and the smaller is the photoelectron current.

The modulation ion trap has four nets. The input net electric potential is equal to the spacecraft body potential. A negative voltage of slightly higher than a hundred volts is applied to the anti-photoelectron net. The photoelectric current from the anti-photoelectron net to the collector can be significant; however, it is not modulated by square voltage pulses applied to the modulating net mounted between the input net and the anti-photoelectron one. An additional (fourth) net is mounted between the modulating net and the anti-photoelectron net to eliminate the capacitive coupling between the modulating net and the collector. Electrons coming into the device are captured by the modulating net. Protons with energy less than the square voltage pulse amplitude, coming into the device, create the modulated ion current on the collector. The alternating component of the ion current is amplified, rectified, and measured, while the unmodulated photoelectron current is rejected. The detector sensitivity limit is now defined by the amplifier noise. The energy separation is realized by varying the voltage applied to the modulating net. If the energy of coming particles is close to 1 keV, the square pulse of +0.1 kV creates almost no modulation; however, the square pulse of +1.5 kV creates the strong modulation.

Usually, an energy spectrum is measured on several levels. It allows constructing an ion energy integral spectrum but leads to significant errors in reconstructing the ion energy distribution. If the modulation voltage varies between two nonzero values, the Faraday cylinder allows constructing an ion energy differential spectrum.

One can divide the directed motion from thermal ones when one measures fluxes coming from several distinct directions (for example, when a spacecraft rotates and plasma characteristics do not change during the rotation period).

The Faraday cylinder drawbacks are the poor angle resolution, the photo-current from nets and walls, and a few minutes needed to obtain an energy spectrum.

The electrostatic analyzer consists of two concentric spherical surfaces (spherical segments or hemispheres with the mean radius *r*) divided by a small interspace. There is the voltage U between spherical surfaces. Suppose there is a balance between electrostatic and centrifugal forces acting the particle coming into the interspace. In that case, such particle passes the interspace and will be registered on the collector. The registered collector current is proportional to the flux of passing particles. Changing the value of voltage (U) allows measuring fluxes of particles having various energies. This way, the device measures a differential (not integral) energy spectrum. The device energy resolution is ~ 10%. The angle of 120° between the plasma flow direction and the normal to the collector is chosen to diminish the photoelectron emission from the collector: solar photons can achieve the collector in case of at least two "lucky" reflections from black spherical surfaces. Neither the energy determination nor the energy resolution depends significantly on the plasma flow incident angle. However, the practical device aperture decreases by 20% for the incident angle of 6°, relatively to the normal incidence. Because a spacecraft usually has a nonzero velocity component (V_{sc}) perpendicular to the plasma flow direction, the aberration arises, and the effective incidence angle is equal to $arctg(V_{sc}/V)$. This effect impedes the registration of low energy protons and establishes the low energy limit to 200 eV. Mounting the device on a rotating spacecraft allows obtaining information about the ion direction of arrival.

The electrostatic analyzer allows obtaining an ion energy differential spectrum and an ion bulk velocity with an accuracy of ~ 10%. Ion concentration estimates depend strongly on the plasma flow incidence angle: minor errors in determining the angle could lead to significant ion concentration estimate errors.

The principle of operation of an induction magnetometer is based on the Faraday induction law: the change of magnetic flux across any circuit creates the electromotive force, which is directly proportional to the magnetic flux value. A typical detector of the induction magnetometer consists of many thin wire coils (usually from 3×10^4 to 6×10^4) on the large magnetic permittivity metal core. The device measures both the magneto-static field and magnetic variations. Usually, the device is mounted on an axially rotating spacecraft, and the detector axis should be orientated perpendicular to the spacecraft rotation axis. In the constant magnetic field, the output voltage is sinusoidal. The sinusoid frequency is equal to the angular velocity of the spacecraft rotation. The amplitude is proportional to the magnetic field component perpendicular to the rotation axis. To determine the field direction, one compares any point of the sinusoid (for example, the passage through zero) to any fixed direction in space (the solar position, for example). In magnetic variations, the sinusoidal output voltage is superimposed by the voltage induced by these variations. The induction magnetometer does not sense the internal magneto-static fields of a spacecraft itself. Therefore, it is useful when flux-gate (saturated core) magnetometers cannot be used without the compensation of spacecraft fields.

The flux-gate magnetometer detector contains two parallel solenoids with wire coils twisted in opposite directions. The high magnetic permittivity metal cores are saturated under the relatively weak magnetic flux density. The solenoids are connected in series and exited by the alternating current with the amplitude sufficient to saturate cores during most of the semi-period. The flip coil is twisted around two solenoids. The flip coil is connected with the tuned amplifier through the capacity and connected to the DC amplifier. The cores are adjusted so that the cycle of magnetization of the first coil is opposite to the cycle of magnetization of the second one. Therefore, the net magnetic flux through the flip coil remains zero in the absence of an external magnetic field. The presence of the external magnetic field component parallel to the detector axis breaks the symmetry of cycles of magnetization, and core polarities will not vary synchronously. The nonzero magnetic flux will pass through the flip coil between two polarity changes. The flip coil output will be two oppositely directed pulses of the amplitude proportional to the external magnetic flux rate of change. After rectifying and smoothing, the DC output signal is proportional to the external magnetic field. The signal is sent to the DC amplifier with the gain factor chosen so that the flip coil field (H_s) will be 10^5 times stronger than the resulting field (H_e). Under such choice, the external field (H_a) is as follows:

$$H_e = 10^{-5} H_s = H_a - H_s \rightarrow H_a = H_s - 10^{-5} H_s \cong H_s$$

It is equal (with the relative accuracy of 10^{-5}) to the field created in the flip coil. One can calculate the external field (H_s) from measurements of the DC through the flip coil. In the scheme with the high gain factor, the result depends on the parameters of the flip coil only and is independent of other elements of the scheme. The coil magnification factor weakly depends on the temperature; the temperature compensation, realized in most flux-gate magnetometers, decreases the temperature coefficient up to the values of ~ 1 nT per 1°C. The flux-gate magnetometer sensitivity is ~ 0.1 nT. The drawback of the device is that the magnetic detector axis is not obligatory to coincide with its geometric axis. The suitable orientation of the detector should be adjusted experimentally: the detector freely rotates up to 180° under calibration measurements. Three detectors orientated in mutually orthogonal directions on a stabilized spacecraft are used for magnetic field vector measurements. The spacecraft orientation should be known with the accuracy of ±1° to determine the magnetic field direction.

There are a few peculiarities of solar wind plasma measurements on spacecraft. The power consumption of magnetometers and plasma devices is small. However, one should change the threshold voltage in detectors to measure the energy spectra of plasma particles. Thus, one needs to apply the voltage from the spacecraft power system to the device. Any onboard electric circuit generates both magneto-static fields and magnetic variations. Such artificial fields should be minimized. Plasma devices measure weak electric currents; therefore, artificial noise is essential for these devices. On the other hand, mounting the magnetometer on the stock at some distance from the spacecraft can complicate the spacecraft orientation system operation and spacecraft thermal regime maintenance. Ensuring a not shielded field of view can demand to mount plasma devices on stocks outside the spacecraft to eliminate the influence of spacecraft construction elements and the spacecraft self-atmosphere on the measurements of low energy particle fluxes. The spacecraft orientation control and the scanning system operation are complicated because the detector should be directed to the Sun, which is seen under the solid angle of only 0.5° from the Earth orbit. The scanning of the Sun is needed sometimes. When using scanning systems, one should control not only the spacecraft orientation but also the detector orientation. If the device is mounted on the stock, one should consider the stock's dynamics and its influence on the spacecraft stability.

An example of reconstruction of solar wind hydrodynamic parameters from electric current measurements

Let us describe the reconstruction procedure as an example of the "Prognoz 7" spacecraft data processing (Ермолаев 2002, 60-66). The spacecraft has launched the orbit with an initial perigee of 472 km and an initial apogee of 202627 km on October 30, 1978. An initial orbit inclination was 64.9°, and an initial orbital period – 5881.1 minutes. The spacecraft was in the solar wind during 50–70% of the orbital period. It was stabilized by self-rotation with a period of about 2 minutes. Due to the Earth orbital motion, the angle between the self-rotation axis and the solar direction changed with a rate of ~ 1° per day; however, the spacecraft orientation system ensured the presence of the spacecraft long axis inside the cone opening ±10° relative to the solar direction.

The plasma spectrometer СКС-04 on the "Prognoz 7" spacecraft measured the integral flux of ions and electrons with energies greater than 100 eV and energy spectra of protons and α-particles.

Two identical detectors measured energy spectra. Each detector consisted of three electrostatic analyzers added (for the particle selection based on its velocity) with the Vin-filter. These were mounted in the plane sequentially so that the axis of the central one was orientated along the long spacecraft axis, which, in turn, was directed to the Sun. The others were shifted to the angles +7.5° and -7.5° relative to the spacecraft long axis correspondingly. An electrostatic analyzer determined the ion energy spectrum in the range 0.23–4.6 keV. Every spectrum was reconstructed from measurements with 24 evenly logarithmically spaced energy steps. Steps were switched every 10.24 seconds simultaneously in all three devices. The Vin-filter turning ensured the selection of ions with the given energy to charge ratio and the given mass to charge one. When the first detector registered the proton energy spectrum, the second detector registered the α-particle energy spectrum and vice-versa.

Three identical Faraday's cylinders measured angles of direction and the integral flux of ions and electrons with energies greater than 100 eV. The conic viewing field of Faraday's cylinder is characterized by the cone opening of 90° and the half-height width of 40°. Three cylinders were separated from each other by the azimuthal angle of 120°, and its axis of viewing field was shifted at the angles of 20° from the long spacecraft axis. The angular diagram of the three cylinders was relatively wide to register the integral flux of solar wind ions and electrons. The comparison of the currents from separated cylinders allows determining the solar wind flux direction.

Special precautions were foreseen to diminish the solar ultraviolet influence on the results of measurements. These included blacking and corrugating metal parts under the influence of the incident or the reflected solar light. Also, protecting nets were mounted above collectors of Faraday's cylinders. Negative voltage -100 V was applied to the net to return photoelectrons to the collector. Nevertheless, photoelectric currents in Faraday's cylinders remained comparable with the solar wind integral flux.

The plasma payload of the "Prognoz 7" spacecraft allows deriving the solar wind ion bulk velocity, the density, and the kinetic temperature only. When registering charge particles of *l*-sort, the current on the Faraday cylinder collector is as follows:

$$I_l = q_l n_l \iiint v f_l F d^3 v$$

Here q_l is the particle charge, n_l is the concentration, v is the particle velocity, f_l is the velocity distribution function of particles of *l*-sort, F is the device parameter. The total current on the collector is as follows:

$$I_s = I_i + I_e + I_\Phi$$

Here I_i and I_e are ion and electron currents, and I_Φ is the photoelectron current.

Measurements on the "Prognoz 7" spacecraft revealed that the photoelectron current was stable and did not depend on the device orientation relative to the solar direction: in magnetospheric regions, where there were no particle fluxes greater than 10^7 cm^{-2} s^{-1}, the photoelectron current was ~ 10^{-9} A or 20% of I_i and varied no more than 10% during the spacecraft passage across the regions (~ 4 days). It allowed estimating the photoelectron current daily mean value with an accuracy of no more than 20%.

For the almost isotropic distribution function of solar wind electrons, $I_e \ll (I_i, I_\Phi)$. The Faraday cylinder angle capacity function varies linearly from unity under the flux incidence angle of 0° to zero under the flux incidence angle of 45°. Therefore, the part (P) of the omnidirectional total flux (P_0) of isotropically distributed particles can be estimated as follows:

$$\frac{P}{P_0} = \frac{2\pi \int_0^{\pi/4} \sin\left(\frac{\pi}{4} - x\right) dx}{2\pi \int_0^{\pi} \sin x \, dx} \le 5 \cdot 10^{2}$$

The Faraday cylinder registered only electrons with energies more significant than the net protecting potential (100 V). Hence, the part of electrons having the Maxwell distribution with the electron temperature of

20 keV (the average solar wind electron temperature at the Earth orbit) was estimated as follows:

$$\xi = 1 - \frac{\int_0^{\sqrt{5}} \exp\left(-\frac{x^2}{2}\right)dx}{\int_0^{\infty} \exp\left(-\frac{x^2}{2}\right)dx} \leq 2.6 \cdot 10^{-2}$$

Therefore, the electron to ion flux ratio was estimated as follows:

$$\frac{I_e}{I_i} = \frac{n_e V_{Te}}{n_i V_i} \xi \frac{P}{P_0} \leq 8 \cdot 10^{-3}$$

Here n_e and n_i are electron and ion concentrations, V_{Te} is the thermal velocity of electrons, V_i is the mass velocity of ions. The total angle characteristic of three Faraday's cylinders varied no more than 20% and was almost independent of the azimuthal angle.

Because the solar wind ion bulk velocity is an order of magnitude higher than its thermal velocities, and the ion motion direction varies no more than 15% around the solar direction, one can suppose F=1 when calculating the solar wind ion flux:

$$P = n_i \iiint v f d^3 v = n_i V_i$$

If one takes into account that 95% of solar wind ions are monovalent ones, the ion flux is calculated as follows:

$$P = n_i V_i = \frac{A}{q}\sum_{j=1}^{3}(I_j - I_{\phi j})$$

Here I_j (and $I_{\Phi j}$) is the current measured by j-Faraday's cylinder (and its photoelectron current), A is the geometric factor of the three cylinders (coefficient). The measured currents I_j were compared with the response function J_j calculated assuming the shifted Maxwell ion velocity distribution and taking into account the device angular characteristic to determine flux incidence angles. The ion flux thermal width was simulated with the help of previously calculated standard tables for various thermal to bulk motion energy ratios. Errors of ion flux calculations were no more than 20% and were stipulated mainly by the uncertainties in estimating photoelectron currents.

Being given the ion bulk velocity and the flux, one can calculate the ion concentration. The ion bulk velocity, the kinetic temperature, and the α-particle abundance one determined from the electrostatic analyzer data under the assumption of the isotropic Maxwell velocity distribution in the reference frame moving with the solar wind flow. In the spacecraft's frame of reference, one should consider the corresponding change of both value and direction of the ion bulk velocity. For the electrostatic analyzer of the high energy resolution ($\Delta E << E$) and the slight angle factor ($<<2kT/m_i V^2$, where m_i is the ion mass, V is the ion bulk velocity, T is the kinetic temperature, and k is the Boltzmann constant), the device reading on k-step is as follows:

$$Y_\kappa(E_\kappa) = CE_\kappa\left(\frac{kT}{m_i}\right)^{3/2} \exp\left[-\frac{\left(\sqrt{E_\kappa}-\sqrt{E_0}\right)^2}{kT}\right] \qquad (3.1)$$

Here C is the concentration, $E_0=m_iV^2/2$ is the kinetic energy of the solar wind flow. One determined the parameters (C, E_0 (or V), and T) minimizing the function of function:

$$S=\sum_{\kappa=1}^{N}(y_\kappa - Y_\kappa)^2$$

Here N was the number of significant points in the energy spectrum; y_k and Y_k were measured and calculated from Eq. (3.1) response values.

Besides the assumption of shifted Maxwell ion velocity distributions, the primary source of uncertainty in determining the velocity, the temperature, and the concentration of ions was “one dimensional” measurements of energy distribution in the situation of the actual unknown angle between the device axis and the ion flow direction. Firstly, actual three-dimensional distributions could be anisotropic (two different kinetic temperatures along and perpendicular to the interplanetary magnetic field). Secondly, it could demonstrate distribution maxima for different ion species at different incident flow angles.

The accuracy of the α-particle abundance determination with the help of the narrowly directed detector depends not only on the device characteristics but also on the flow characteristics such as the solar wind flow thermal width and incidence angles of proton and α-particle fluxes. Modelling revealed that the α-particle abundance estimate could differ from its actual value no more than 2–3 times depending on the incidence angles of protons and α-particles. The estimates are valid for the detector energy resolution ~ 6%, the detector angular resolution ~ 5%, the difference between α-particle and proton incidence angles no more than 2–3°, and typical thermal widths of solar wind proton flows and α-particle ones.

In principle, fluxes of ions heavier than α-particles could be reconstructed from the solar orientated electrostatic analyzer data. The proton peak (E/q = 0.6 keV) and the α-particle peak (E/q = 1.2 keV) could be selected in an energy spectrum. Due to the detector's high sensitivity operating in such a mode, the proton peak could be in punning. Still, one could separate peaks with amplitudes enhancing 0.6 KeV more than two times, which were not secondary peaks of protons or α-particles and could be interpreted as manifestations of heavy ions.

Velocity distributions of solar wind components

The solar wind plasma is non-equilibrium. The Maxwell distributions shifted on the solar wind flow bulk velocity can only roughly describe the velocity distributions. Measured (Marsch et al. 1982, 35-51) (Marsch et al. 1982, 52-72) velocity distributions demonstrated both the difference between proton and electron temperatures and the thermal anisotropy ($T_{\|}$, $T_{\perp}$) relative to the interplanetary magnetic field direction. These were approximated by shifted bi-Maxwell distributions as follows:

$$f_l \sim \exp\left[-\frac{m_l}{2T_{l\|}}\left(\vec{v}_l - \vec{V}_l\right)_{\|}^2 - \frac{m_l}{2T_{l\perp}}\left(\vec{v}_l - \vec{V}_l\right)_{\perp}^2\right], l = i, e$$

Proton velocity distributions demonstrated two maxima most of the time and could be interpreted as two mutually penetrating flows. In the main flow characterizing by the bulk motion energy ~ 500 eV and the thermal dispersion ~ 10 eV, the temperature perpendicular to the interplanetary magnetic field was higher than the temperature along the interplanetary magnetic field. The additional "fast" flow (an additional energy maximum ~ 10–20 keV) demonstrated the opposite temperature anisotropy. The concentration in the group of "fast" protons was ~ 10^{-3}–10^{-2} cm^{-3}. At the same time, its energy density in disturbed conditions was enhanced up to 10^{-10} erg cm^{-3}, which was comparable with the thermal energy density of the main solar wind flow. The main flow moved almost radially from the

Sun, while the "fast" proton flow was directed from the Sun along the interplanetary magnetic field. The relative velocity of mutually penetrating flows was of the order of the local Alfven speed and decreased with the heliocentric distance. Reported were cases when mutually penetrating flows were diminished after the magnetic field changed the direction from radial to tangential. The origin of "fast" protons remains unclear up to the moment.

Velocity distributions of protons and α-particles were obtained along the "Helios 2" spacecraft trajectory between Mercury and the Earth. From 20% to 30% of all analyzed spectra demonstrated the second maximum containing 10–20% of all protons (Marsch et al. 1982, 35-51). Three-dimensional velocity distributions of α-particles also differed from thermal ones. The thermal anisotropy increased with the heliocentric distance increase. Supra-thermal α-particles, as well as α-particle velocity distributions with two maxima, were often observed. A maximum in the α-particle velocity distribution, as a rule, was situated between two maxima in the proton velocity distribution. In the mutually penetrating proton flows, α-particles in their velocity distribution maximum moved faster than protons of the main proton velocity distribution maximum but slower than protons of the secondary velocity distribution maximum. The bulk velocity of α-particles was higher than the bulk velocity of the main proton population on the value of the local Alfven speed. Incidence angles of protons oscillated with the amplitude up to 10° around the solar direction in azimuthal and meridional planes; the corresponding interplanetary magnetic field oscillated with the amplitude up to 45° around the solar direction. α-particles moved almost radially from the Sun (Bame et al. 1975, 463-473) (Marsch et al. 1982, 52-72).

Obtaining electron velocity distributions on the "IMP 7" spacecraft revealed the existence of the "electron core" with the temperature ~ 10^5 °K, and the "electron halo" with the temperature ~ 7×10^5 °K and the concentration ~ 5% of the total electron concentration (Веселовский, 1984, 98-103). The "electron halo" moved from the Sun faster than the "electron core": the relative velocity was ~ 800 km s^{-1}. The "electron core" moved from the Sun slower than protons: the relative velocity was ~ 50 km s^{-1}. The "electron halo" transferred almost all heat: the "electron core" deposit to the heat flow was negligible. Noticed was the correlation between the "electron core" velocity relative to the solar wind bulk velocity and the "electron halo" temperature anisotropy $(T_{\|}/T_{\perp})_H$. Electron temperatures and concentrations were reconstructed from the data obtained on the "Mariner 10" spacecraft along its flight trajectory to Mercury at heliocentric distances of 0.85–0.45 AU. With the increase of the heliocentric distance, the "electron core" temperature decreased as the power law with the index -0.3±0.04 while the electron concentration – as the power law with the index -2.5±0.2. The "electron halo" temperature was independent of the heliocentric distance. Two shifted bi-Maxwell distributions with different temperatures and concentrations can describe the "electron core" and the "electron halo". Parameters of the "electron core" are $n_c = 10.0\pm4.8$ cm^{-3}, $V = 425\pm73$ km s^{-1}, $T_c = (1.25\pm0.29)\times10^5$ °K, $(T_{\|}/T_{\perp})_c = 1.08\pm0.08$. Parameters of the "electron halo" are $n_H = 0.57\pm0.23$ cm^{-3}, $\Delta V_H = 689\pm369$ km s^{-1}, $T_H = (6.9\pm1.1)\times10^5$ °K, $(T_{\|}/T_{\perp})_H = 1.22\pm0.18$. The "Helios 1-2" spacecraft data revealed another peculiarity of the electron velocity distribution. It was narrow and clear asymmetry along the interplanetary magnetic field direction: the electron phase space density for energies greater than 100 eV was several orders of magnitude higher in the direction from the Sun than in the direction to the Sun (Pilipp et al. 1981, 241-249) (Pilipp et al. 1987, 1075-1092) (Rosenbauer et al. 1977, 561-580). For energies greater than 600 eV, the asymmetry was often narrowed to several degrees. In such strongly asymmetric electron velocity distributions, the main heat flow was transferred by the tail of electron velocity distribution, not the "electron halo".

The fundamental reason for non-Maxwell velocity distributions in the solar wind is the increase of the particle free path with the particle velocity: slower particles collide more often than faster ones. It increases the relaxation time to the Maxwell velocity distribution.

Solar wind turbulence

Cosmic plasmas often are turbulent – significant irregular fluctuations of its physical parameters take place. The solar wind is not the exception: such fluctuations are discovered for all spatial and temporal scales. The distinguishing feature of turbulence is its local non-reproductive character. Nevertheless, the turbulence demonstrates a few "universal" statistical properties in a few cases. For example, the turbulence in the isotropic incompressible liquid (where the only fluctuating parameter is the liquid velocity) can be imagined as the continuous fragmentation of the velocity fluctuation (δv) spatial scale. Inertia forces ensure the fragmentation. Assumed is that the fluctuation energy is pumped to the system on a spatial scale comparable to the system scale. This energy is transferred to smaller scales (l) at a constant rate. At the possible minor scale, the transferred energy is converted into heat; hence, the temperature increases. On the spatial scale of velocity fluctuations much smaller than the spatial system scale but much larger than the scale on which minor velocity fluctuations are destroyed, the constant rate of the energy fluctuation transfer ($\varepsilon \sim (\delta v)^2/(l/\delta v) \sim (\delta v)^3/l$) means that $\delta v \sim (\varepsilon l)^{1/3}$. As the wavenumber $k \sim l^{-1}$, the spectrum of velocity fluctuations $(\delta v)^2/k \sim k^{-5/3}$ demonstrates an example of such "universal" statistical characteristics (Kolmogorov 1941, 301-305).

The magnetized solar wind turbulence differs radically from the turbulence in the isotropic incompressible neutral liquid: the magnetic field destroys the turbulence isotropy. It leads to a longer correlation length along the field than the one perpendicular to the field $l_{||} > l_{\perp}$ (Matthaeus et al. 1990, 20673-20683). Roughly speaking, one can suppose that turbulent vortexes are shorter in the direction perpendicular to the field than their scales along the field direction. It means that the turbulent energy of wave vectors at great angles relative to the magnetic field ($k_{\perp} > k_{||}$) dominates so that the wave vector distribution is anisotropic. Spacecraft measurements of both the wave vector anisotropy and corresponding spectra in the solar wind are not trivial. Such measurements allow obtaining time data series along the spacecraft trajectory. The Fourier transformation gives the frequency spectrum but not the wave vector spectrum. Usually, the solar wind velocity is much more than both characteristic plasma velocities (the local Alfven speed is ~ 50 km s^{-1} at the Earth orbit) and the spacecraft velocity. It allows using the Taylor hypothesis (Taylor 1938, 467-490) (Perri and Balogh 2010, 17102-17105) to transform the frequency f, measured in the spacecraft frame of reference, to the wave vector component along the solar wind velocity (V) direction in the reference frame moving with the solar wind plasma element $k = 2\pi f/V$. If the solar wind plasma moves along the magnetic field, one measures fluctuations of the wave vector component parallel to the magnetic field direction ($k_{||}$). If the solar wind velocity is perpendicular to the magnetic field, one measures fluctuations of the wave vector component perpendicular to the magnetic field direction ($k_{\perp}$). One may use the Taylor hypothesis only in the absence of whistlers propagating along the magnetic field with a phase velocity greater than the solar wind velocity. “Helios 1-2” spacecraft observed the whistler noise enhancement during the flight to the Sun. Magnetic field oscillations were measured with the help of two three-axis magnetometers in frequency ranges between 0 and 4 Hz and between 4 Hz and 2.2 kHz. Magnetometers were mounted on 2.8 m length and 4.6 m length stocks correspondingly and had low noise levels. The data were averaged over 1.125 s time intervals. Observed was the slowly varying (on the time scale longer than 10 min) background noise with the typical spectral density 10^{-2}-10^{-3} nT per $Hz^{1/2}$ on 7 Hz and 10^{-4} nT per $Hz^{1/2}$ on 70 Hz. The spectral density depended on the frequency as the power law with the index values between -1.4 and -2.0 under the thickest value -1.7. This density varied on an order (or two) of magnitude during a solar rotation. Closer to the Sun, the spectral density was increased more than an order of magnitude, and the spectrum became steeply. Electromagnetic oscillations demonstrated a frequency lower than 220 Hz at the Earth orbit 99% of the time. The oscillations with a frequency lower than 470 Hz were registered at 0.5 AU 94% of the time (Beinroth and Neubauer 1981, 7755-7760). One should remember that the magnetic fluctuation energy density is several orders of magnitude lower than the total magnetic energy density. Magnetic fluctuation spectra measured on the "Ulysses" spacecraft at the heliocentric distance interval (1.38–1.93) AU in the fast high heliolatitude solar wind revealed that the spectrum of transverse (to the magnetic field direction) oscillations decreased according to the power-law with the index -5/3. The spectrum of oscillations along the magnetic field direction contained less energy, and its power-law index was -2 (Wicks et al. 2010, L31-L35). The maximal frequency (the external scale of turbulence) is $\sim 10^{-4}$ Hz in the solar wind at the Earth orbit. At the large scale ($f < 10^{-4}$ Hz in the spacecraft reference frame), oscillations were isotropic: both parallel and perpendicular oscillation spectra were of the same order and decreased linearly when the frequency increased (Bruno and Carbone 2005, 4-185). Mangeney, with colleagues (Mangeney et al. 1991, 327-356), assumed that the turbulent cascade could be developed at the heliocentric distance r where the vortex origin time is shorter than the wind expansion time r/V. Estimates at the

Earth orbit under the solar wind velocity 600 km s^{-1} gave ~ 70 hours (f_0 ~ 10^{-4} Hz). From the Taylor hypothesis under the typical velocity oscillation amplitude of $\delta v \sim 25$ km s^{-1} per Hz$^{1/2}$, the spatial scale of oscillations at the frequency $f_0 = 10^{-4}$ Hz can be estimated as $l \sim 600$ km s$^{-1}/f_0 \sim 6\times10^6$ km. In the range (10^{-4}, 10^{-1}) Hz, observed was the spectrum splitting: turbulent magnetic oscillations developed mainly in the perpendicular (to the magnetic field direction) plane and had the Kolmogorov spectrum $E(k_\perp) \sim k_\perp^{-5/3}$. In contrast, the parallel (to the magnetic field direction) oscillations demonstrated the spectrum $E(k_{||}) \sim k_{||}^{-2}$. Oscillations along the magnetic field direction were much weaker than oscillations in the plane perpendicular to the magnetic field. It testified for the Alfven fluctuation dominance in this frequency range. When processing experimental data, one should consider that the solar wind plasma element passes near the spacecraft only once and never returns. Hence, the average direction of fluctuating magnetic field on the spatial scale of many correlation lengths measured during a certain time interval not obligatory corresponds to the averaged magnetic field real direction. Such a procedure rather determines the local magnetic field direction for every fluctuation, and then fluctuations with the same local magnetic field directions are averaged. Studied were fluctuations in the frequency range between 2.4×10^{-5} Hz and 1.3×10^{-2} Hz of both the magnetic field value and its components at the heliocentric distance interval (0.29–1) AU in the ecliptic (the "Helios 1-2" spacecraft data). The magnetic field value spectra became steeper when the heliocentric distance increased, while spectra of field components decreased more slowly than magnetic field value spectra (Denskat and Neubauer 1982, 2215-2223). The spectrum of solar wind velocity fluctuations approximated as power-law with the index -3/2 was much flatter than the magnetic one (Grappin et al. 1991, 416-426). For frequencies lower than 10^{-5} Hz, the power of field-aligned velocity oscillations was two orders of magnitude higher than the power of transverse velocity oscillations. In comparison, at higher frequencies, both powers had comparable values. For frequencies lower than 10^{-1} Hz, the spectrum of plasma density fluctuations was Kolmogorov's one (Celnikier et al. 1983, 293-298). The rate of energy transfer across the spectrum varied from ~ 0.1 kJ kg^{-1} s^{-1} in the fast high latitude solar wind to ~ 10 kJ kg^{-1} s^{-1} in the slow solar wind at the Earth orbit (McComas et al. 2003, 1517-1521) (Sorriso-Valvo et al. 2007, 115001-115001). Compressible fluctuations contained only a small part of the turbulence energy. At higher frequencies, the spectrum became more flat (decreased linearly with the frequency) at first but then returned to the steep curve again. At the Earth orbit, the magneto-hydrodynamic turbulence cascade is finished near frequencies ~ (0.1–0.3) Hz – the estimates of the ion cyclotron frequency in the spacecraft reference frame. At such frequencies, various kinetic effects supply energy to the turbulent cascade or extract it from there. In particular, the spherically symmetric corona expansion generates fluctuations via mirror or hose instabilities taking place under high ion plasma beta values (higher than 3) (Bale et al. 2009, 211101-211102). However, the exact partition of energy between hydrodynamic and kinetic degrees of freedom remains the subject of discussion.

Statistics of solar wind velocity, density, temperature, magnetic field, and derivative parameters

The solar wind plasma parameters in interplanetary space are studied in great detail by numerous direct measurements on spacecraft. We follow the paper (Veselovsky et al. 2010, 113-128), based on the mean hourly parameters (http://omniweb.gsfc.nasa.gov/form/dx1.html). We present the mean values of solar wind parameters and their variations for the time encompassing four solar cycles (20–23) with maximums in 1969, 1980, 1989, 2000, and minimums in 1964, 1976, 1986, 1996, and 2007. Over the time under study, the limits of the Wolf number variations were from 0 to 302. Some regular solar wind variations between the cycles and relatively small cyclic variations were discovered previously. In the majority of cases, all their values are of the order of ten percent.

The data set of solar wind parameters for 1963–2007 includes about 4×10^4 hourly mean counts, about one-third of them being gaps: the set is incomplete. This fact substantially complicates the data analysis and the correct statistical interpretation of the long but incomplete data series. Short-covering by the data (less than 50%) occurs in the initial period of scattered investigations in 1963–1972 and then in 1983–1993 when regular measurements in the interplanetary space were conducted only by the "IMP 8" spacecraf. In addition, measurements on high-apogee satellites can give distorted information about conditions in the solar wind due to the proximity to the magnetosphere boundary. The most severe problems with the quality of plasma measurements emerge during solar eruptive events and radiation damages from proton events when reliable data about the solar wind density and the temperature can be unavailable for several days. The histograms of solar wind plasma basic parameters were constructed using the database mentioned above. The methods of construction are standard (Deeming 1975, 137-158). Essential statistical characteristics of solar wind parameters are presented in Table 3-1.

Table 3-1 Statistics of basic physical parameters of the solar wind at the Earth orbit in 1963–2007 (adapted from Veselovsky et al. 2010, 117)

1	2	3	4	5	6	7	8	9	10	11
V, kms^{-1}	265699	156	1200	420	430	1.25	1.26	0.42	0.32	p
T, $10^{3o}K$	227883	2.9	7000	85	83	2.3	2.43	0.14	0.31	l
D, cm^{-3}	244625	0.1	118.0	5.3	5.4	2.0	2.0	0.01	0.59	l
He/p	163052	10^{-3}	0.4	0.04	0.03	2.0	1.7	0.79	1.4	a
B, nT	271938	0.4	55.8	5.9	6.0	1.55	1.48	0.12	0.85	l
B_{xy}, nT	271938	0.14	52.0	4.6	4.5	1.74	1.55	0.78	2.8	l
B_x, nT	271938	40.1	29.4	0	0	3.9	-	-	-	t
B_y, nT	271938	38.8	46.1	0	0	4.3	-	-	-	t
B_z, nT	271938	46.3	36.8	0	0	3.1	2.09	0.01	9.1	n

Note: 1 is a parameter symbol, 2 is the statistics (the number of hourly mean counts), 3 is the minimum value, 4 is the maximum value, 5 is the distribution median, 6 is the first distribution moment (the arithmetic mean value), 7 is the normalized second distribution moment (the dispersion), 8 is the normalized standard deviation, 9 is the normalized third distribution moment (the asymmetry coefficient), 10 is the normalized fourth distribution moment (the excess parameter), 11 is the distribution form (p=plane, l=lognormal, a=asymmetric, t=two peaks, n=normal).

There are some indications in the literature that most solar wind parameters are close to the lognormal distribution (Hartlep et al. 2000, 5135-5139) (Burlaga and Lazarus 2000, 2357-2364). Therefore, statistical distributions in Fig. 3-1 and their approximations are presented in the logarithmic scale. Statistical distributions of the density *n* and the temperature *T* (Figs. 3-1a and 3-1b) have relatively small third and fourth moments (the asymmetry and the excess). These distributions are well approximated by the lognormal law so that ~ 95% of the entire statistics, which includes, respectively, 244625 and 227883 mean hourly values, lies within the limits of 2-σ interval (see Table 3-1).

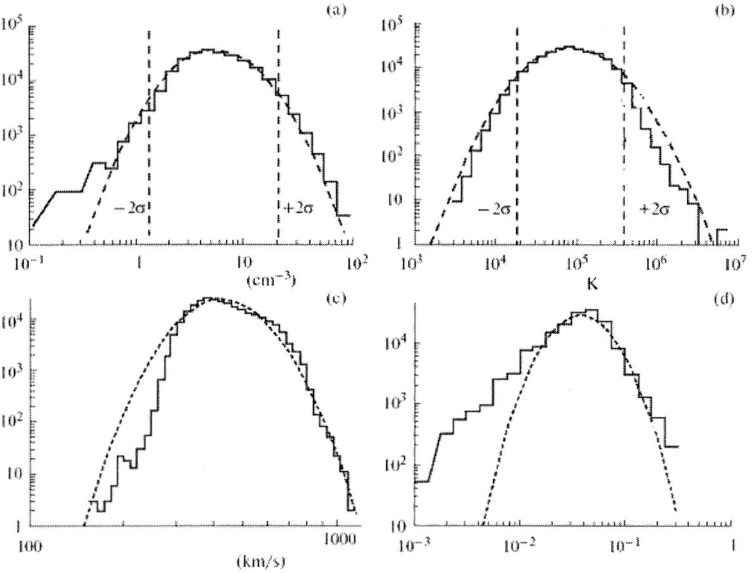

Fig. 3-1 Statistical distributions of the solar wind parameters represented in the logarithmic scale for the time interval from 1963 to 2007 for (a) proton density, (b) proton temperature, (c) velocity, (d) ratio *He/p*; approximations by the lognormal distribution are shown as a dashed line; thin vertical dashed lines limit the zone of good approximation (adapted from Veselovsky et al. 2010, 118)

Distributions of the solar wind velocity and the abundance of helium nuclei are not such and look completely different (Figs. 3-1c and 3-1d). A specific feature of the density distribution within the 2-σ interval is that it well correlates with the solar wind velocity. The density distribution outside this interval has less steep wings (Fig. 3-1a). The most noticeable distinction at the smallest values $n < 0.4$ cm^{-3} is partially associated with long-lasting phenomena, when the solar wind almost disappeared and $n < 0.1$ cm^{-3} (Lazarus 2000, 2172-2173) (Crooker et al. 2000, 3769-3772). Also, it is associated with short-term rarefaction regions lasting a few hours (Usmanov et al. 2005, https://doi.org/10.1029/2004JA010699) (Janardhan et al. 2008, https://doi.org/10.1029/2007JA012608) and with the fact that in the used database rounding of numbers is performed to a first significant digit in the lower order. The largest values of density, $n > 20$ cm^{-3}, arise in shock waves. Enormous densities (up to $D \sim 100$ cm^{-3} and higher) are observed in ejected filaments. It should be noted that the duration of such time intervals in the solar wind is short, about one hour; therefore, their contribution on average is small (Bothmer and Schwenn 1995, 1127-1132) (Burlaga et al. 1998, 277-285) (Dal Lago et al. 2001, 451-455). The density D of the solar wind plasma was calculated, taking into account the proton density and corrections for helium ions with the mass number 4 using the formula $D = n(1 + 4He/p)$. Here n is the proton density, and He/p is the helium content relative to protons.

The proton temperature distribution is well approximated by the lognormal law in the region from $\sim 10^4$ °K up to $\sim 4\times10^5$ °K (Fig. 3-1b). Inside this interval, the temperature correlates with the solar wind velocity (Lopez and Freeman 1986, 1701-1705) (Lopez 1987, 11189-11194) so that fast flows turn out to be hotter. Compared with this model approximation, a deficit of very cold and hot fluxes is observed outside the 2-σ interval. The wings of the observed distribution are somewhat narrower, and its variance is lower (2.3) compared to the model value (2.43). It was demonstrated that low temperatures ($T < 1.5\times10^3$ °K) could be observed for several days in succession, being accompanied by low-velocity values (from 200 to 300 km s^{-1}) (Freeman and Lopez 1985, 9885-9894). In addition, low temperatures can be met in coronal mass ejections and ejected filaments (solar prominences) (Richardson and Cane 1995, 23397-23412). One can think that the above peculiarities of the statistical distribution in the wings are caused by the interaction between solar wind flows and are due to the equalization of their velocities and temperatures. The deficit at high temperatures > 4×10^5 °K is better pronounced than at low temperatures (Fig. 3-1b). This circumstance can be related to the fact that the fast wind is less dense. Therefore, any changes are the most substantial and most noticeable just here under the interaction of flows. The fast flows from high heliolatitudes can impart their momentum to slower flows from low heliolatitudes when colliding with them on the curved current sheet (Mc Comas et al. 2003, 1517-1521). The long tail in the statistical distribution at the highest temperatures > 2×10^6 °K does not go into the linear relation between the velocity and the temperature and exceeds it (Skoug et al. 2004, https://doi.org/10.1029/2004JA010494). Such extreme temperatures arise at plasma braking and heating on shock waves from coronal mass ejections. Thus, the thermodynamic properties of the solar wind plasma to a large degree are associated with its velocity.

The velocity distribution is remarkably different from the lognormal one (Fig. 3-1c), and its third and fourth moments are significant. The value of the second moment is close to the model variance. The most probable velocity value of ~ 380 km s^{-1} is lower than its median value of ~ 420 km s^{-1}. In the region from ~ 300 km s^{-1} to ~ 700 km s^{-1}, the statistical distribution of probability has a very wide maximum containing more than 95% of all mean hourly counts (their total number in the database is 265699).

The statistical distribution of the relative abundance of helium ions *He/p* in the solar wind is shown in Fig. 3-1d. The histogram form with broad and asymmetric wings is distinct from both normal and lognormal distributions. The third and fourth moments are relatively large. Hourly mean values of the ratio *He/p* lie within limits from ~ 10^{-3} to ~ 0.3 at the mean value of 0.04. The statistics of the composition data is relatively poor. Before 1970, there were no data on the solar wind composition, and regular measurements have been carried out since 1998. As a result, one can say that the coverage of less than 50% took place over the period

1963–2007. Gaps are usually filled with a mean value of 0.04. Such an approximation is somewhat rough. The long tail stretching out to the side of low values down to ~ 10^{-3} is partially cut when rounding data to the first decimal digit. The solar wind ion composition formation is highly complicated and poorly studied. It is reduced to the separation and mixing of ion components in the source region and during the propagation in the interplanetary space. In one way or another, the observed upper limit in the degree of enrichment of the solar wind by helium ions gives unambiguous evidence that mixing imposes specific constraints. It can be considered as yet another important indication of the turbulent character of the solar wind formation. On average, the abundance of helium correlates with the solar wind velocity (Aellig et al. 2001, 2767-2770). Lesser values correspond to the slow solar wind, while larger values are associated with the fast wind (Borrini et al. 1981, 4565-4573). Ultimately large values > 0.1 correspond to coronal mass ejections with an ejected solar prominence inside them (Aellig et al. 2001, 2767-2770).

The statistical distributions of the interplanetary magnetic field value and its components were considered in the lognormal approximation (Burlaga 2001, 15917-15927). The attention was drawn to the existence of significant deviations from this approximation in the form of big third and fourth moments in the data averaged over 3 hours (Feynman and

Ruzmaikin 1994, 17645-17651). More complicated approximations were considered, for example, with fixed mean values and normally distributed components (Hartlep et al. 2000, 5135-5139). The anisotropic character of magnetic fluctuations in general, with comparable amplitudes along and across the average magnetic field direction, was noted (Bieber et al. 1993, 3585-3603). Fig. 3-2a shows the statistical distribution of probability for the magnetic field magnitude B in the logarithmic scale. It is close to the lognormal distribution within the 2-σ corridor to the side of low values and 1.5-σ to high values relative to the mean value equal to 6 nT. More than 95% of the data fell within this corridor. Less than 5% of the total counts remain in the distribution wings. The fraction of the values < 2 nT is less than 1%. It is known that so-called magnetic holes in the solar wind can last for several hours, and they are not connected with magnetic field reversals (Zurbuchen et al. 2001, 16001-16010). Enormous values $B > 13$ nT are observed in 4% of all events, and they are accompanied by the substantial variability of solar wind parameters. Coronal mass ejections and compression regions before them can cause this excess over the lognormal distribution (Burlaga and Szabo 1999, 137-140) (Burlaga et al. 1987, 5725-5734) (Owens and Cargill 2002, https://doi.org/10.1029/2001JA000238) (Owens et al. 2005, https://doi.org/10.1029/2001JA000238) (Bothmer and Schwenn 1995, 1127-1132) (Dal Lago et al. 2001, 451-455).

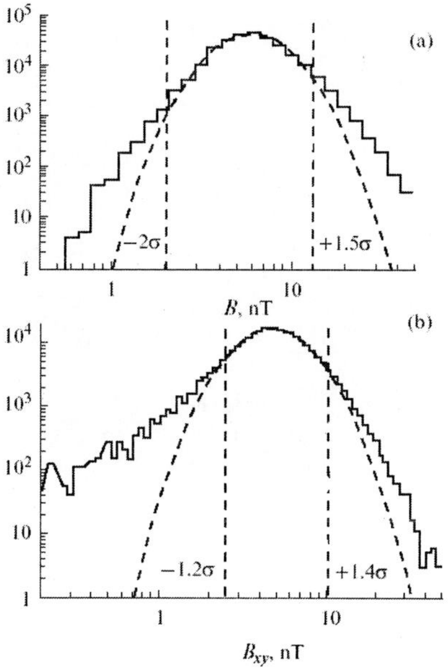

Fig. 3-2 Statistical distributions of the interplanetary magnetic field from 1963 to 2007 in the logarithmic scale for (a) the magnitude of B and (b) the projection of the interplanetary magnetic field vector onto the ecliptic B_{xy}; the best approximation by the lognormal distribution is shown by the dashed line; thin vertical dashed lines limit the zone of good approximation (adapted from Veselovsky et al. 2010, 121)

The statistical distribution of the magnetic field vector component $\boldsymbol{B}_{xy}$ (the vector sum $\boldsymbol{B}_{xy} = \boldsymbol{B}_x + \boldsymbol{B}_y$) is shown in Fig. 3-3b. It is close to the lognormal distribution with the mean value of 4.6 nT in the narrow corridor from -1.2 to +1.4 standard deviations, including about 80% of the total statistics. Fig. 3-3 shows two-dimensional statistical distributions (B_x,B_y) of magnetic field vector components in the solar-ecliptic coordinate system with axis X directed to the Sun and axis Z directed to the north.

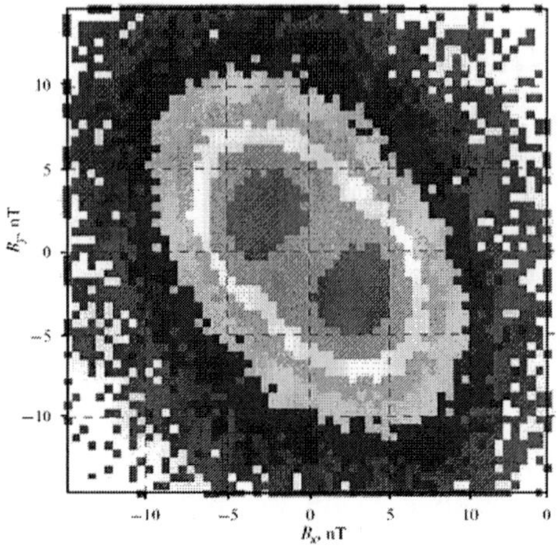

Fig. 3-3 Statistical distributions of B_x and B_y components of the interplanetary magnetic field in the geocentric solar-ecliptic coordinate system; this indicates the predominant orientation of the field vector along the direction of ~135° relative to the radial direction to the Sun (positive values of B_x) (adapted from Veselovsky et al. 2010, 122)

One can see a range at an angle of ~ -45° (or ~ 135°) (Luhmann et al. 1993, 5559-5572). It is interesting to note that this direction is typical not only for mean values of the field but in many cases for strong fields exceeding 15 nT. The angle varies within wide limits, and, as a result, a large dispersion in the angle value is observed, which is well seen in Fig. 3-3. Large amplitude Alfven waves additionally make a substantial contribution to these estimates of the interplanetary magnetic field dispersion (Belcher and Davis 1971, 3534-3563) (Tsurutani and Gonzales 1987, 405-412) (Tsurutani et al. 1995, 21717-21733). The wavelength spectrum ranges up to several million kilometres and corresponds to periods of the order of one hour and more prolonged. These waves are most clearly observed in the high-speed streams on the back front to the solar rotation. In the compression regions on leading edges of such streams (so-called corotating interaction regions), the amplitude of these waves reaches ~ 20 nT. Their polarization can be different. In particular, they can produce the north-south magnetic field component B_z. The statistical distribution of this quantity is shown in Fig. 3-4.

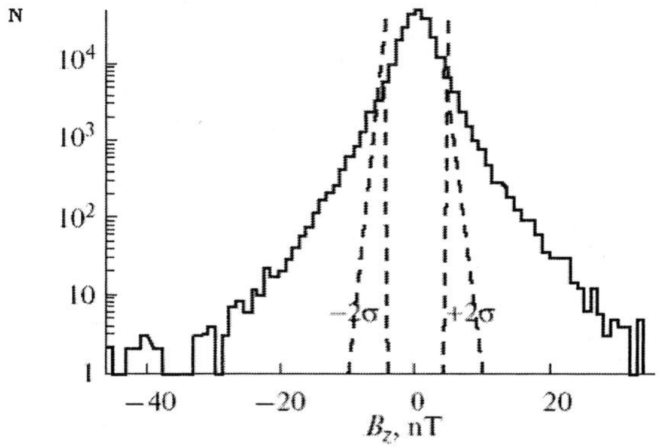

Fig. 3-4 The statistical distribution of B_z component in the solar wind for the time interval from 1963 to 2007; thin vertical dashed lines indicate the interval of the best approximation by the normal distribution (adapted from Veselovsky et al. 2010, 123)

Its third moment is zero, and it is well approximated by the normal law with a zero mean value in the 2-σ corridor, i.e. within ±4.18 nT. However, the fourth moment is positive and quite significant. More than 13% of all mean hourly values are concentrated in the distribution wings. The most significant field intensities are generated under the compression in interaction regions of corotating and transient streams (Bothmer and Schwenn 1995, 1127-1132). Tables 3-2, 3-3 and 3-4 present some parameters characterizing the properties of the interplanetary medium at the Earth's orbit and calculated using the average values presented above. The proton velocity closely approximates the mass velocity of helium ions in most cases, and, in calculations, they were put equal with an accuracy of ten kilometres per second (Borrini et al. 1982, 7370-7378) (Ogilvie et al. 1982, 7363-7369).

Table 3-2 Mean values of the fluxes of energy and mass at the Earth orbit in 1963–2007 (adapted from Veselovsky et al. 2010, 120)

Physical quantity	Mean value
The mass flux density (m_pDV)	4.0×10^{-16} g cm^{-2}s^{-1}
The flux density of kinetic energy ($0.5m_pDV^3$)	0.44 erg cm^{-2}s^{-1}
The flux density of potential energy ($0.5m_pV_g^2DV$, V_g=618 km s^{-1})	1.7 erg cm^{-2}s^{-1}
The flux density of proton enthalpy ($2.5nTV$)	6.8×10^{-1} erg cm^{-2}s^{-1}
The flux density of magnetic energy ($VB^2/8\pi$)	6.1×10^{-3} erg cm^{-2}s^{-1}

Table 3-3 Average conditions in the solar wind plasma at the Earth orbit in 1963–2007 (adapted from Veselovsky et al. 2010, 124)

Physical quantity	Mean value
Alfvenic velocity ($B(4\pi nm_p)^{-1/2}$)	56.8 km s^{-1}
The sound velocity for protons ($(5T/3m_p)^{1/2}$)	33.7 km s^{-1}
The Alfven-Mach number	7.7
The Mach number for protons	13.0
The ratio of gas and magnetic pressures (plasma β)	0.43
Coulomb's time for electron collisions ($\sim10^{-2}T_e^{3/2}n^{-1}$)	$\sim9.9\times10^4$ s
Coulomb's time for proton collisions ($\sim0.6T_p^{3/2}n^{-1}$)	$\sim2.9\times10^6$ s

Table 3-4 Basic plasma characteristics calculated for average solar wind conditions in 1963–2007 (adapted from Veselovsky et al. 2010, 124)

Physical quantity	Mean value
The plasma frequency $((4\pi ne^2)^{1/2}m_e^{-1/2})$	1.3×10^5 s^{-1}
The electron cyclotron frequency $(eB(m_e c)^{-1})$	1.04×10^3 s^{-1}
The proton cyclotron frequency $(eB(m_p c)^{-1})$	0.56 s^{-1}
The mean thermal velocity for protons $((3T/m_p)^{1/2})$	45.8 km s^{-1}
The Larmor radius for protons $(V_p m_p c(eB)^{-1})$	81.8 km

The lognormal distribution is multiplicative, i.e. the product of two lognormally distributed quantities is also lognormal. Therefore, logarithmically mean values are presented in the Tables. Table 3-2 presents the values of flux densities for energy and mass. Thermal and magnetic energy fluxes are two orders of magnitude less than the solar wind kinetic energy flux at the Earth orbit. The thermal flux carried by solar wind protons along the magnetic field is directed from the Sun. It is much less than the solar wind kinetic energy flux, the enthalpy flux, and the thermal flux carried by solar wind electrons. Relative errors of flux estimates are no better than ~ 20%. Flux values vary a dozen percent inside a solar activity cycle and from cycle to cycle. The most significant contribution to the energy flux is its kinetic and potential components. The fluxes of both the thermal energy (enthalpy) and the magnetic energy in the solar wind are smaller by two orders of magnitude. Let us remind you that the solar radiation energy flux is the main one. It exceeds all other kinds of dissipation by several orders of magnitude (Veselovsky et al. 1999, 485-486). Table 3-3 presents some mean values of plasma and electrodynamic characteristics necessary for various physical estimates. Notice that here the density value is used without taking α-particles into account, which can introduce an error of the order of 10%. The lack of sufficiently complete data on electron temperatures adversely affects the sound speed estimates, the Mach number, the total gas pressure, and the gas to magnetic pressure ratio. The electron temperature T_e varies in the wide range from ~ 5×10^4 °K up to ~ 10^6 °K. It correlates very weakly with the proton temperature (Newbury et al. 1998, 9553-9566), which testifies to the incomplete thermodynamic equilibrium in the solar wind. In fast solar wind streams, the proton temperature is approximately twice higher than the electron temperature, which can be due to features of forming such streams in the corona. Electrons turn out to be hotter than protons approximately by a factor of four in the slow and dense solar wind. According to these incomplete data (Table 3-1), the mean temperature of electrons and protons is about ~ 1.4×10^5 °K and 8.5×10^4 °K, respectively. Therefore, neglecting the contribution of electrons, one can underestimate the sound speed approximately twice, and accordingly, sonic Mach numbers will be overestimated. The same can be said about some other parameters, which include this quantity. One can deal only with sufficiently rough estimates. On average, the solar wind flow at the Earth's orbit is supersonic and superalfvenic: the bow shock near the Earth exists virtually always, even under disturbed conditions, when it can significantly approach and even more significant move away from its nominal position calculated using the average values of solar wind parameters (Spreiter et al. 1966, 223-250).

There are indications that rare events of the solar wind subalfvenic flow at the Earth's orbit can be associated not with coronal mass ejections but rather with flows having a low density (< 0.3 cm^{-3}) and (only in some cases) an intense interplanetary magnetic field (> 10 nT) (Usmanov et al., 2005, https://doi.org/10.1029/2004JA010699). Extremely high Alfven-Mach number values are observed inside magnetic holes (Zurbuchen et al. 2001, 16001-16010).

The fact that solar wind parameters in many manifestations so definitely indicate the lognormal distribution can testify indirectly to the concept of multiple multiplicative transformations of local characteristics at intermitting random intensifications and attenuations of waves, compressions and rarefactions of irregularities in turbulent processes in the solar wind. At the moment, this scenario of development seems to be the most realistic. On the other hand, it is known, according to the central limit theorem of the probability theory, the normal (Gaussian) distribution originates as a result of the summation of many independent random quantities. The lognormal distribution is generated at the multiplication of random quantities. However, as far as it goes, the lognormal distribution is not a sufficient argument for conclusions about these processes' predominantly random and irregular character. Deterministic phenomena (their algorithm being sufficiently complicated and unknown) can have virtually no differences with random processes for which there is no algorithm in principle. Mimicry of the complex and the simple frequently has its source in the lack of knowledge about the physical nature of the processes involved and about dimensionless scales implicated, along with their geometrical realization in space and time.

As far back as 1962, the "Mariner 2" spacecraft first observed high-speed solar wind streams (twice faster than the background) in the ecliptic. The tendency to recur with the solar equatorial rotation period allowed interpreting high-speed streams as long-lived spatial structures. Measurements on "Helios 1-2" spacecraft in ecliptic in 1975 testified to the existence of fully developed high-speed streams already at heliocentric distances of 70 solar radii (Веселовский, 1984, 47-48). There were eight radial oppositions of the "Helios 1" spacecraft and the "Helios 2" spacecraft when both spacecraft were situated along a single radial heliocentric ray. During such short opposition intervals, the opportunity existed to measure radial gradients of solar wind parameters (Schwenn et al. 1981, 126-130). It was discovered that the velocity in the slow wind (with velocities lower than 400 km s^{-1}) slightly increased with the heliocentric distance while the proton temperature decreased almost adiabatically there. The velocity in the fast wind remained almost constant, and the proton temperature decreased twice slowly as in the slow wind. The proton temperature was twice higher than the electron one in the fast wind, while electrons were four times hotter than protons in the slow wind.

The interplanetary magnetic field heliocentric distance dependence was measured on a few spacecraft between Mercury and Jupiter orbits during various periods. Most of the data testified to the square-law decrease of the radial component with the heliocentric distance. The tangential component decreased linearly.

According to the "Ulysses" spacecraft observations, the fast tenuous solar wind from solar polar caps dominates everywhere besides the slow solar wind band in the heliolatitude interval of 15–25° near the ecliptic. In the fast polar solar wind, the radial magnetic field component does not depend on the heliolatitude. Spatial structures of balanced total (thermal plus magnetic) pressure exist in the polar solar wind as well as micro-streams (daily local velocity enhancements of ~ 50 km s^{-1} above the background value of ~ 750 km s^{-1}) and mutually penetrating flows (Smith et al. 1993, 2327-2330) (Balogh et al. 1995, 1007-1010) (Balogh and Forsyth 1998, 48-51) (Mc Comas et al. 2003, 1517-1521).

Discontinuities in the solar wind

Spacecraft often register "sharp" changes of solar wind parameters on spatial scales ~ 4×10^3 km (~ 3×10^{-5} AU). Such phenomena are interpreted as magneto-hydrodynamic discontinuities. The conservation of mass, momentum and energy across the discontinuity along with the continuity of both the electric field tangential component and the magnetic field normal component creates the linkages connecting solar wind parameters on opposite discontinuity sides (Баранов, Краснобаев, 1977, 92-99) as follows:

$$\{B_n\} = 0, B_n\{\vec{V}_\tau\} = m\left\{\frac{\vec{B}_\tau}{\rho}\right\}, \{m\} = 0 \qquad (3.2)$$

$$\left\{m\vec{V}_\tau + \left(p_\parallel - p_\perp - \frac{B^2}{4\pi}\right)\frac{B_n\vec{B}_\tau}{B^2}\right\} = 0,$$

$$\left\{p_\perp + \frac{m^2}{\rho} + \frac{B_\tau^2}{8\pi} + \frac{B_n^2}{B^2}(p_\parallel - p_\perp)\right\} = 0 \qquad (3.3)$$

$$m\left\{\varepsilon+\frac{m^2}{2\rho^2}+\frac{p_\perp}{\rho}+\frac{V_\tau^2}{2}+\frac{B_n^2}{B^2}\frac{p_\parallel-p_\perp}{\rho}+\frac{B_\tau^2}{4\pi\rho}\right\}+$$
$$B_n\left\{\frac{p_\parallel-p_\perp-B^2/4\pi}{B^2}\left(\vec{B}_\tau\cdot\vec{V}_\tau\right)\right\}=0 \tag{3.4}$$

Because the discontinuity “thickness” is much less than the solar wind character spatial scale, the discontinuity surface can be considered as the plane (with the normal $\boldsymbol{n}$) moving with the velocity $\boldsymbol{S}$. Here $p_\parallel$ ($p_\perp$) is the field-aligned (the transverse) pressure, ρ is the plasma density, V is the plasma velocity, $m = \rho(V_n–S_n)$. Indexes n and τ denotes normal and tangential to the discontinuity plane components, ε is the internal energy in volume unit ($\varepsilon = p_\perp + p_\parallel/2$). The symbol $\{X\} = X_1–X_2$ denotes the difference between the value before the discontinuity (index “1”) and the value after the discontinuity (index “2”). It is supposed that $\{E_\tau\} = 0$ can be replaced by the second of Eqs. (3.2) when $\boldsymbol{E} = -(1/c)[\boldsymbol{V}\times\boldsymbol{B}]$. Eq. (3.3) describes the momentum conservation across the discontinuity in tangential and normal projections to the discontinuity plane. Eqs. (3.2)-(3.4) form the closed system only in the case of $p_\parallel = p_\perp = p$ and known S_n.

Adding $\boldsymbol{n}B_n m\{1/\rho\}$ to left-hand and right-hand parts of the second of Eqs. (3.2), one obtains:

$$B_n\left\{\vec{V}\right\}=m\left\{\frac{\vec{B}}{\rho}\right\}$$

$\boldsymbol{B}_1$, $\boldsymbol{B}_2$, and $\{\boldsymbol{V}\}$ are coplanar vectors when $B_n \neq 0$ and $m \neq 0$. This means:

$$\left(\vec{V}_1 - \vec{V}_2\right)\cdot\left[\vec{B}_1 \times \vec{B}_2\right] = 0 \tag{3.5}$$

If $B_n = 0$ and $m \neq 0$, then $\boldsymbol{B}_1 \| \boldsymbol{B}_2$. If $B_n \neq 0$ and $m = 0$, then $\boldsymbol{V}_1 = \boldsymbol{V}_2$. If $B_n = m = 0$, then $\boldsymbol{B}_1$, $\boldsymbol{B}_2$, and $(\boldsymbol{V}_1 - \boldsymbol{V}_2)$ lie on the discontinuity plane. Eq. (3.5) contains all the above cases.

Let us consider the case $B_n = m = 0$. The discontinuity of this type is called the tangential discontinuity: there is no mass flow across the discontinuity, and the magnetic field is tangential to the discontinuity plane. One obtains from the second of Eqs. (3.3):

$$\left\{p_\perp + \frac{B_\tau^2}{8\pi}\right\} = 0 \tag{3.6}$$

The tangential discontinuity allows an arbitrary jump of $p_\|$, ρ, V_τ, but magnetic field and transverse pressure jumps are connected via Eq. (3.6). The velocity $\boldsymbol{S}$ (only S_n has the physical sense) of the discontinuity plane motion and its orientation (the normal $\boldsymbol{n}$) can be determined from the most reliably measured parameters – the solar wind velocity and the magnetic field. Because the magnetic field is parallel to the tangential discontinuity plane, one obtains:

$$\vec{n} = \frac{\vec{B}_1 \times \vec{B}_2}{\left|\vec{B}_1 \times \vec{B}_2\right|}, S_n = \vec{S}\cdot\vec{n} = \vec{V}_1 \cdot \frac{\vec{B}_1 \times \vec{B}_2}{\left|\vec{B}_1 \times \vec{B}_2\right|} \tag{3.7}$$

The plasma velocity (similar to the magnetic field) parallels the tangential discontinuity plane in the discontinuity reference frame. Inside this limitation, values and orientations of both the velocity and the magnetic field can vary arbitrarily. In case of the magnetic field jump across the tangential discontinuity, the plasma density and proton and electron temperatures must change to preserve the pressure conservation across the discontinuity. Plasma and field variations for many discontinuities observed in the solar wind meet these requirements. However, the pressure balance has not been checked strictly because of the incomplete information about electron temperatures.

Let us consider the case $B_n \neq 0$ and $m = 0$. The discontinuity of this type is called the contact discontinuity: there is no plasma flow across the discontinuity, but the normal (to the discontinuity plane) magnetic field component exists – the second of Eqs. (3.2) along with $m = 0$ show that $\boldsymbol{V}_1 = \boldsymbol{V}_2 = \boldsymbol{S}$. An arbitrary jump of plasma density across the discontinuity is allowed, while the other plasma parameters are linked through the relationships:

$$\left\{\frac{p_{\|} - p_{\perp} - B^2/(4\pi)}{B^2}\right\} = 0, \left\{p_{\perp} + \frac{B_\tau^2}{8\pi} + \frac{B_n^2}{B^2}\left(p_{\|} - p_{\perp}\right)\right\} = 0$$

When $p_{\|} = p_{\perp} = p$, the above relationships mean that the magnetic field and the pressure are continuous across the contact discontinuity. In the case of

$\boldsymbol{B}_1 \neq \boldsymbol{B}_2$, the pressure cannot be isotropic. Because $\boldsymbol{B}_1$, $\boldsymbol{B}_2$, $\boldsymbol{n}$ are coplanar vectors, one obtains:

$$\vec{n} = -\frac{\left[\left(\vec{B}_1 - \vec{B}_2\right)\times\left[\vec{B}_1 \times \vec{B}_2\right]\right]}{\left|\left[\left(\vec{B}_1 - \vec{B}_2\right)\times\left[\vec{B}_1 \times \vec{B}_2\right]\right]\right|}, D_n = \vec{n}\cdot\vec{V}_1$$

Let us consider the case $B_n = 0$ and $m \neq 0$. The case corresponds to the shock wave moving perpendicular to the magnetic field direction. One obtains from the second of Eqs. (3.2):

$$\frac{\vec{B}_1}{\rho_1} = \frac{\vec{B}_2}{\rho_2}, \left(\vec{B}_1 \parallel \vec{B}_2\right)$$

One obtains from the first of Eqs. (3.2):

$$\vec{V}_{1\tau} = \vec{V}_{2\tau}$$

Because the tangential velocity component is continuous and $m \neq 0$, then:

$$\vec{n} = \frac{\vec{V}_2 - \vec{V}_1}{\left|\vec{V}_2 - \vec{V}_1\right|}, S_n = \frac{\rho_1\vec{V}_1 - \rho_2\vec{V}_2}{\rho_1 - \rho_2}\cdot\vec{n}$$

The expression for S_n is the consequence of the last of Eqs. (3.2). Because of $B_n = 0$ for this type of shock wave, the relationship takes place:

$$\vec{B}_1 \cdot \left(\vec{V}_2 - \vec{V}_1\right) = \vec{B}_2 \cdot \left(\vec{V}_2 - \vec{V}_1\right)$$

Changes of plasma parameters in the process of the shock wave passage are so quick that they look sudden when registering with the help of onboard plasma detectors. In the shock wave front region, the plasma motion kinetic energy is partly dissipated into the heat. The thickness of the wavefront region is of the order of the proton cyclotron radius. The first information on the shock wavefront spatial configuration was derived from the statistics of shock wavefront local orientations registered on a spacecraft or several spacecraft closely situated in space. However, amplitudes of magnetic fluctuations after the front are enhanced more than an order of magnitude. It created significant uncertainties in determining the direction of normal. The shape of the shock wavefront could be determined from observations of the front on several spacecraft separated each other in space on the distance comparable with the spatial scale of the shock wave as a whole. There are fast and slow shock waves. The fast shock wave is the result of the evolution of fast magneto-sonic waves. The magnetic field in such a shock wave is inclined to the shock wave surface and increases across the surface. The slow shock wave is the result of the evolution of slow magneto-sonic waves. The magnetic field in such a shock is inclined from the shock wave surface and decreases across the surface.

Let us consider the case $B_n \neq 0$ and $m \neq 0$ when the vectors $\boldsymbol{B}_1$, $\boldsymbol{B}_2$, $\boldsymbol{n}$ are coplanar. The case corresponds to the inclined shock wave. From the second of Eqs (3.2) and the first of Eqs (3.3), one obtains:

$$B_n m\left(\vec{V}_{\tau 1}-\vec{V}_{\tau 2}\right)=m^2\left(\frac{\vec{B}_{\tau 1}}{\rho_1}-\frac{\vec{B}_{\tau 2}}{\rho_2}\right)=$$
$$B_n^2\left[\vec{B}_{\tau 1}\left(\frac{1}{4\pi}+\frac{p_{\perp 1}-p_{\parallel 1}}{B_1^2}\right)-\vec{B}_{\tau 2}\left(\frac{1}{4\pi}+\frac{p_{\perp 2}-p_{\parallel 2}}{B_2^2}\right)\right]$$

From where:

$$\vec{B}_{\tau 1}\left[\frac{m^2}{\rho_1}-\frac{B_n^2}{B_1^2}\left(\frac{B_1^2}{4\pi}-p_{\parallel 1}+p_{\perp 1}\right)\right]=$$
$$\vec{B}_{\tau 2}\left[\frac{m^2}{\rho_2}-\frac{B_n^2}{B_2^2}\left(\frac{B_2^2}{4\pi}-p_{\parallel 2}+p_{\perp 2}\right)\right] \tag{3.8}$$

It is clear from Eq. (3.8) that $\boldsymbol{B}_{\tau 1}||\boldsymbol{B}_{\tau 2}$ when the expressions inside square brackets are nonzero, that means:

$$\left(V_{nk}-S_n\right)^2=\frac{B_n^2}{B_k^2}\frac{B_k^2/(4\pi)+p_{\perp k}-p_{\parallel k}}{\rho_k},(k=1,2) \tag{3.9}$$

The coplanar theorem works in this case. For $k = 1$, the shock is called the inclusion shock, for $k = 2$ – the tripping one. For $p_{\parallel} \neq p_{\perp}$, the coplanar theorem works even under $\{\rho\} = 0$. When $\rho_1 \neq \rho_2$ one obtains:

$$\vec{n}=-\frac{\left[\left(\vec{B}_1-\vec{B}_2\right)\times\left[\vec{B}_1\times\vec{B}_2\right]\right]}{\left|\left[\left(\vec{B}_1-\vec{B}_2\right)\times\left[\vec{B}_1\times\vec{B}_2\right]\right]\right|},S_n=\vec{n}\cdot\frac{\rho_2\vec{V}_2-\rho_1\vec{V}_1}{\rho_2-\rho_1}$$

The rotational discontinuity means that $B_n \neq 0$ and $m \neq 0$ and Eqs. (3.9) are valid. The plasma velocity relative to the discontinuity is defined by Eq. (3.9); therefore, one obtains from the last of Eqs. (3.2):

$$\rho_1\left[1-\frac{4\pi\left(p_{\parallel 1}-p_{\perp 1}\right)}{B_1^2}\right]=\rho_2\left[1-\frac{4\pi\left(p_{\parallel 2}-p_{\perp 2}\right)}{B_2^2}\right]$$

When $p_{\parallel} = p_{\perp} = p$, the rotational discontinuity means $\{\rho\} = 0$. In anisotropic case ($p_{\parallel} \neq p_{\perp}$), the rotational discontinuity with $\{\rho\} \neq 0$ may exist. One obtains the necessary condition for the rotational discontinuity existence by substituting Eq. (3.7) into the second of Eqs. (3.3):

$$p_{\perp 1} - p_{\perp 2} + \frac{B_1^2 - B_2^2}{8\pi} = 0$$

When $p_{\parallel} = p_{\perp} = p$, the plasma density does not change across the discontinuity. It leads to the continuity of the thermal pressure and the continuity of the magnetic pressure. The magnetic field rotates around the normal to the rotational discontinuity front. It means that the rotational discontinuity in the isotropic plasma is, in essence, the sharp "bend" propagating along the magnetic field line with the local Alfven velocity. Rotational discontinuities are observed at the Earth orbit approximately once an hour. The discontinuity thickness is of the order of ten proton cyclotron radii. In the anisotropic case, both the plasma density and the magnetic field can jump on the rotational discontinuity.

Summary of Chapter 3

In-situ plasma measurements on spacecraft established that protons are the most abundant ions in the solar wind. The abundance of the others is, as a rule, much less: the concentration of α-particles, for example, is estimated as 4–5% of the proton one. The abundance of α-particles correlates with the solar wind bulk velocity. The slow wind contains fewer α-particles, the fast one – more. The temperature of α-particles is usually higher than the proton one. At the same time, its bulk velocity is close to the proton one (the difference is of the order of the local Alfven velocity).

Relative errors of bulk velocity measurements are estimated as 2–3%, of concentrations and temperatures of protons and α-particles – as 20–30%, of the abundance of α-particles – as 200–300%.

The average value of the solar wind bulk velocity is estimated as ~ 380 km s^{-1}. In the interval between 10^4 K and 4×10^5 K, the proton temperature correlates with the solar wind bulk velocity. Thermal and magnetic energy fluxes are two orders of magnitude less than the solar wind kinetic energy flux at the Earth orbit. The thermal flux carried by solar wind protons along the magnetic field is directed from the Sun. It is much less than the solar wind kinetic energy flux, the enthalpy flux, and the thermal flux carried by solar wind electrons. Relative errors of flux estimates in the solar wind are no better than ~ 20%. Solar wind fluxes vary a dozen percent inside a solar activity cycle and between cycles.

Mainly the proton velocity distribution demonstrates two maxima and can be interpreted as the manifestation of two mutually penetrating flows. The transverse temperature is higher than the temperature along the magnetic field direction in the main flow (with the proton directed motion energy of 500 eV and thermal dispersion of ~ 10 eV). There is the opposite thermal anisotropy in the additional "fast" flow (with the proton directed motion energy of ~ 10–20 keV). The "fast" proton concentration is ~ 10^{-3}–10^{-2} cm^{-3}, while in disturbed conditions, its energy density is ~ 10^{-10} erg cm^{-3} and comparable with the thermal energy density of the main proton flow. The main flow moves radially from the Sun while the weak "fast" flow is directed along the interplanetary magnetic field. The relative velocity of the two flows is of the order of the local Alfven speed. Mutually penetrating flows usually are followed by thermal anisotropy ratio enhancements. Velocity distributions of α-particles also demonstrate two maxima. Usually, the maximum of α-particles is situated between two proton maxima so that the velocity of α-particles is higher than the velocity of the main proton group. The velocity difference is of the order of the local Alfven velocity. Electron velocity distributions demonstrate the existence of the core with the temperature of ~ 10^5 °K and the halo with the temperature of ~ 7×10^5 °K. The halo moves ~ 800 km s^{-1} faster than the core. The latter moves ~ 50 km s^{-1} slower than solar wind protons. The halo carries almost all electron thermal flux. Electron velocity distributions are narrowed in the transverse (to the magnetic field) direction sometimes. In such cases, the electron thermal flux is carried by high-velocity tail electrons of the core, not by halo ones.

The typical value of the interplanetary magnetic field at the Earth orbit is 5–7 nT. The magnetic field line in the ecliptic at the Earth orbit is inclined on the angle ~ -45° (or ~ 135°) relative to the radial direction. The interplanetary magnetic field radial component decreases quadratically when the heliocentric distance increases, while the tangential one decreases linearly.

Spacecraft register recurrent high-speed (moving twice faster than the background) streams close to the ecliptic. The proton temperature is twice higher than the electron one in the streams, while electrons are four times hotter than protons in the slow solar wind. In the solar wind with velocities lower than 400 km s^{-1}, the proton temperature decreases almost adiabatically. In high-speed streams, the proton temperature decreases only slightly.

The solar wind almost disappears sometimes – proton concentrations lower than 0.1 cm^{-3} were registered. The highest proton concentrations (up to 100 cm^{-3} and more) are met in shock waves and prominences. The prominences are relatively cold. Low proton temperatures are also met in coronal mass ejections. The highest proton temperatures are registered in shock waves. The electron temperature is much less variable because of the high electron thermal conductivity.

According to the "Ulysses" spacecraft observations, the fast tenuous solar wind from solar polar caps dominates everywhere besides the slow solar wind band in the heliolatitude interval of 15–25° near the ecliptic. In the fast polar solar wind, the radial magnetic field component does not depend on the heliolatitude. Spatial structures of balanced total (thermal plus magnetic) pressure exist in the polar solar wind along with micro-streams (daily local velocity enhancements of ~ 50 km s^{-1} above the background value of ~ 750 km s^{-1}) and mutually penetrating flows.

For frequencies lower than 10^{-4} Hz, the solar wind turbulence is isotropic. The external scale of the turbulent cascade is close to 10^{-4} Hz at the Earth orbit: for higher frequencies, the turbulence becomes anisotropic – the correlation length along the magnetic field direction is longer than the corresponding transverse length. The turbulent cascade develops mainly in transverse (to the magnetic field) directions. In the frequency range between 10^{-4} Hz and 10^{-1} Hz, Alfven's fluctuations dominate, while electron concentration fluctuations demonstrate the Kolmogorov spectrum.

Compressive fluctuations contain only a minor part of the solar wind turbulent energy. The rate of the turbulent energy transfer across the spectrum varies between two points. The lower ~ 0.1 kJ kg^{-1} s^{-1} is measured in the fast polar solar wind; the faster ~ 10 kJ kg^{-1} s^{-1} is measured in the slow solar wind at the Earth orbit.

CHAPTER 4

THEORETICAL CONSIDERATIONS OF THE ACCELERATION REGION

In this chapter, we state the theoretical foundations of solar wind acceleration.

At first, we present the kinetic description of the solar wind as the evaporation of the coronal material. In the hot solar corona, the electron thermal velocity is an order of magnitude higher than the solar escape velocity. In contrast, the proton thermal velocity is a few times below the escape velocity. Therefore, the solar gravitation confines protons but cannot confine electrons. The Coulomb attraction between protons and electrons creates the polarization electric field in the corona that impedes electron's outcome from the solar gravitation pit without accompanying protons. In such an approach, the solar wind phenomenon is a manifestation of the significant difference between electron and proton masses under the equality of its electric charges. However, mathematical constructions applied for the kinetic modelling of the phenomenon are too rigid: only the spherically symmetric solar wind in the radial magnetic field can be modeled.

Magneto-hydrodynamics looks more flexible to model the spatially inhomogeneous magnetized solar wind in its acceleration region. Considered is the solar rotation influence on the acceleration process. Coronal roots of recurrent high-speed streams and coronal mass ejections are also considered.

At the end of the chapter, we estimate the polarization electric field in the solar wind acceleration region.

Evaporation of the corona

The density of any atmosphere falls when the distance from the center of gravitation increases; the solar corona is not an exception. The coronal density decrease (as the function of the height h above the solar surface) is derived from the equalization of the gravitation force to the pressure difference dP at altitudes h and $h+dh$:

$$dP = -\rho(h)g_s(h)dh = -n(h)mg_s(h)dh \tag{4.1}$$

(ρ is the density, g_s is the acceleration of gravity, n is the coronal gas concentration, m is the average mass of coronal atoms) taking into account the ideal gas law:

$$P(h) = n(h)kT(h) \rightarrow dP(h) = kT(h)dn(h) + kn(h)dT(h) \tag{4.2}$$

Substituting Eq. (4.1) into Eq. (4.2), one obtains:

$$n(h) = n(h_0)\exp\left\{-\int_{h_0}^{h}\frac{dh}{H(h)} - k\left[T(h) - T(h_0)\right]\right\}$$

Here

$$H(h) = \frac{kT(h)}{m(h)g_s(h)}$$ - is the scale of height

Particularly in the isothermal corona ($T(h)$=*const*), the concentration decreases exponentially with the altitude:

$$n(h) = n(h_0)\exp\left(-\int_{h_0}^{h}\frac{dh}{H}\right) = n(h_0)\exp\left(-\frac{h-h_0}{H}\right)$$

Here $H = \dfrac{kT}{mg_s}$

In the almost fully ionized plasma of the solar corona, the primary particle interactions are the Rutherford scatterings of electrons on ions as well as electrons on electrons and ions on ions. Other elementary processes take place either with photon emissions (the bremsstrahlung in electron-ion collisions is an example) or neutral particle participations (the electron impact ionization of atoms and the charge exchange between ions and atoms are the examples). Every scattering leads to a turn of the particle trajectory. In plasma, scatterings occur mainly on long distances between particles of the order of the Debye length r_D so that deflection angles are small and particle trajectories are smooth (almost straight) lines. The lines slowly change their direction under the action of numerous but very weak momentums due to long-distance interactions. The distance of shortest approach (~ r_D) almost coincides with the impact parameter so that the free path is as follows (Арцимович, 1978, 176):

$$l = \frac{1}{4\pi n}\left(\frac{m_1 v^2}{q_1 q_2}\right)^2 \frac{1}{L_K} \qquad (4.3)$$

Here q_1 (and m_1) is the charge (and the mass) of a test particle, q_2 is the charge of a scattering center, n is the concentration of scattering centers, and L_K is the Coulomb logarithm (variates from 10 to 20). The free path to scale height ratio is known as the Knudsen number (Kn). According to direct observations on spacecraft, the solar wind is neither a case where collisions dominate ($Kn \ll 1$) nor a case of collision-less plasma ($Kn \gg 1$). Such a medium is described by the Boltzmann equation, which is the integrodifferential equation for the distribution $f(\boldsymbol{r},\boldsymbol{v},t)$ of the rarefied gas:

$$\frac{\partial f}{\partial t} + \vec{v}\frac{\partial f}{\partial \vec{r}} + \frac{\vec{F}}{m}\frac{\partial f}{\partial \vec{v}} = \frac{1}{Kn}\int dV_1$$

$$\iint d\Omega\sigma(\theta, \vec{v} - \vec{v}_1)\,|\,\vec{v} - \vec{v}_1\,|\,[f(\vec{r},\vec{v}',t)f_1(\vec{r},\vec{v}_1',t) - f(\vec{r},\vec{v},t)f_1(\vec{r},\vec{v}_1,t)]$$

(4.4)

Here $\boldsymbol{v}$ and $\boldsymbol{v}_1$ are particle velocities before the collision, $\boldsymbol{v}'$ and $\boldsymbol{v}'_1$ are velocities of the same pair of particles after the collision (connected to $\boldsymbol{v}$ and $\boldsymbol{v}_1$ via momentum and energy conservation laws), $d\sigma = \sigma(\theta, \boldsymbol{v}-\boldsymbol{v}_1)d\Omega$ is the differential cross-section of scattering into a solid angle $d\Omega$ (θ is the angle between vectors $\boldsymbol{v}_1$-$\boldsymbol{v}$ and $\boldsymbol{v}'_1$-$\boldsymbol{v}'$). Values of $\boldsymbol{v}'$ and $\boldsymbol{v}'_1$ are uniquely defined by values of $\boldsymbol{v}$, $\boldsymbol{v}_1$, θ. The addendums $\boldsymbol{v}\partial f/\partial \boldsymbol{r}$ and $(\boldsymbol{F}/m)\partial f/\partial \boldsymbol{v}$ defines distribution variations caused by the entry and the exit of particles in/from an element $d\boldsymbol{r}d\boldsymbol{v}$ of the phase space due to the action of the external force $\boldsymbol{F}$. A particle velocity changes abruptly in the collision, and the particle leaves the specified interval $d\boldsymbol{v}$. However, there are also collisions after which a particle with the velocity outside the interval $d\boldsymbol{v}$ comes to this interval. The right-hand side of Eq. (4.4) (the so-called Boltzmann's integral) is the difference between the number of particles contained in an elementary volume $d\boldsymbol{r}d\boldsymbol{v}$ of the phase space at the moment t but will not be found in the volume $d\boldsymbol{r}^*d\boldsymbol{v}^*$ ($\boldsymbol{r}^*=\boldsymbol{r}+\boldsymbol{v}dt$, $\boldsymbol{v}^*=\boldsymbol{v}+\boldsymbol{F}dt/m$) at the moment $t^*=t+dt$ and the number of particles penetrated to the volume $d\boldsymbol{r}^*d\boldsymbol{v}^*$ during dt but were absent in this volume at the moment t. Collisions in plasma are weak, and particles change their motion directions very slowly. Therefore, one can consider only binary collisions of a particle with the other particles, which means that the number of collisions is proportional to ff_1. General solutions of Eq. (4.4) can be constructed in exceptional cases only. The description becomes simpler when analyzing the two extremes. In the first extreme, $Kn \ll 1$, the right-hand side of Eq. (4.4) is much more than its left-hand side, which puts zero. In the second one, $Kn \gg 1$, collisions are neglected; the right-hand side is zero, and the Boltzmann equation degenerates into the Vlasov one. The boundary (so-called exobase) between the two extremes is situated at the heliocentric distance where $Kn = 1$, i.e. where the free path is equal to the scale height in the isothermal corona. The effective cross-section of ion-electron collisions

$$\sigma = \frac{1}{nl}$$

is estimated in the solar corona as $\sigma \cong 10^{-17}$ cm^2 (van de Hulst 1953, 280). It allows estimating the electron concentration at the exobase ($Kn = 1$, $l = H$) as follows:

$$n = \frac{1}{\sigma H} = \frac{1}{\sigma}\frac{m_p G M_s}{kTR} \approx 10^6\, cm^{-3}$$

Here G is the gravitational constant, m_p is the proton mass, R is the solar radius, M_s is the solar mass. So the exobase could be situated at a heliocentric distance somewhere between 2 and 3 solar radii.

All models of the coronal evaporation (so-called exospheric models) always are constructed by joining the two extremes at the exobase: $Kn \ll$

1 (frequent collisions) below the exobase and $Kn >> 1$ (the absence of collisions) above the exobase. All exospheric models do not take into account solar rotation. All exospheric models are spherically symmetric with the radial magnetic field given as follows:

$$B_\theta = 0, B_\varnothing = 0, B_r = \pm\frac{B_0}{r^2}, \ (B_0 = const) \tag{4.5}$$

If the constant B_0 is chosen positive in one hemisphere, it must be chosen negative in the opposite hemisphere to preserve the solenoidal magnetic field described by Eq. (4.5). In this case, the only possible electric current is the azimuthal surface electric current in the equatorial plane.

For $Kn << 1$ between the coronal base and the exobase, Eq. (4.4) degenerates into the equation:

$$\int d\vec{V}_1 \iint d\Omega\sigma(\theta, \vec{v} - \vec{v}_1) | \vec{v} - \vec{v}_1 | [f(\vec{r}, \vec{v}', t) f_1(\vec{r}, \vec{v}_1', t) -$$

$$f(\vec{r}, \vec{v}, t) f_1(\vec{r}, \vec{v}_1, t)] = 0 \tag{4.6}$$

According to Boltzmann's H-theorem, the thermodynamic equilibrium solution to Eq. (4.6) is the local Maxwell velocity distribution:

$$f_\alpha = n_l \left(\frac{m_\alpha}{2\pi k T_\alpha} \right)^{3/2} \exp\left[-\frac{m_{\alpha l}}{2kT_\alpha} \left(\vec{v}_\alpha - \vec{V}_\alpha \right)^2 \right], \alpha = i, e$$

The concentration, the bulk velocity, and the temperature (n, V, T) are functions of space and time slowly varying on the spatial scale of the order of the free path and the temporal scale of the order of the free path time. It is assumed that the free path across the magnetic field is much less than the cyclotron radius; therefore, the magnetic field between the coronal base and the exobase is supposed not too strong.

Thermal velocities of electrons and protons for coronal temperatures are presented in Table 4-1.

Table 4-1 Thermal velocities of electrons and protons in the corona (adapted from Pisanko and Yakovlev 2020, 1050)

T (°K)	$V_e=(2kT/m_e)^{1/2}$ (km s^{-1})	$V_p=(2kT/m_p)^{1/2}$ (km s^{-1})
10^6	5504	128
1.5 10^6	6741	157
2.0 10^6	7784	181
2.5 10^6	8703	203
3.0 10^6	9534	222

The solar escape velocities at various heliocentric distances are presented in Table 4-2.

Table 4-2 Escape velocities at various heliocentric distances (adapted from Pisanko and Yakovlev 2020, 1050)

r (in solar radii R)	V (in km s^{-1})
1	618
2	437
3	357
4	309
5	276

It is clear from the Tables that the electron thermal velocity is almost an order of magnitude higher than the solar escape velocity. In contrast, the proton thermal velocity is several times lower than the solar escape velocity. It means that the solar gravitation confines coronal protons but cannot confine electrons. However, an electron withdrawal from the solar gravitation pit is impeded by the polarization electric field E_r, whose value can be roughly estimated from the following considerations (Pannekoek 1922, 107-118) (Rosseland 1924, 720-729). Let us denote the electric force to gravitation force ratio for a proton as follows:

$$\varsigma = \frac{eE_r}{m_p g_s} \tag{4.7}$$

(e is an elementary electric charge). The equilibrium of forces in the isothermal corona for protons with Maxwell's distribution is as follows:

$$m_p g_s - eE_r = (1-\varsigma) m_p g_s = kT_p \frac{d \ln n_p}{dh} \tag{4.8}$$

For electrons with Maxwell's distribution, it is as follows:

$$m_p g_s + eE_r \cong \varsigma m_p g_s = kT_e \frac{d \ln n_e}{dh} \tag{4.9}$$

The sign of approximate equality in Eq. (4.9) is the consequence of the condition $m_e/m_p << 1$. A sufficient number of collisions ($Kn << 1$) maintains almost equal electron and proton temperatures, while electron and proton concentrations should be almost equal to prevent significant charge separation. It means that the right-hand sides of Eq. (4.8) and Eq.

(4.9) are equal to each other, but this is possible only in the case when $\zeta \cong 0.5$. From Eq. (4.7), one obtains:

$$E_r = \frac{m_p g_s}{2e} = \frac{m_p GM_s}{2er^2} = \frac{m_p}{2e}\frac{d}{dr}\left(-\frac{GM_s}{r}\right) \tag{4.10}$$

Introducing the electric potential:

$$E_r = -\frac{d\varphi_E}{dr} \tag{4.11}$$

and equating right-hand sides of Eq. (4.10) and Eq. (4.11), one obtains:

$$\varphi_E = \frac{m_p}{2e}\frac{GM_s}{r} \tag{4.12}$$

Let us consider the collision-less ($Kn >> 1$) motion of solar wind particles above the exobase. All particles move from the Sun; the absence of collisions eliminates particles scattered to the Sun. The velocity distribution in such a wind obeys the Vlasov equation:

$$\left\{\frac{\partial}{\partial t} + \vec{v}\frac{\partial}{\partial \vec{r}} + \frac{e}{m_\alpha}\left(\vec{g} + \vec{E} + \frac{1}{c}\left[\vec{v} \times \vec{B}\right]\frac{\partial}{\partial \vec{v}}\right)\right\} f_\alpha(\vec{r}, \vec{v}, t) = 0 \tag{4.13}$$

Here α is the index denoting the solar wind component (protons or electrons). In the collision-less plasma above the exobase ($Kn >> 1$), the cyclotron radius is assumed to be much smaller than the spatial character scale of the solar wind, so this spatial character scale is assumed to increase with the heliocentric distance faster than the cyclotron radius does. In such a case, a particle motion in slowly varying electric and magnetic fields can be described by the superposition of the "fast" cyclotron rotation around the magnetic field line and the "slow" drift of the cyclotron rotation center (driving center), while Eq. (4.13) can be reduced to the collision-less kinetic equation in drift approximation (Лифшиц, Питаевский, 1979, 308-311) which (in the stationary case) degenerates into the equation:

$$v_r \frac{\partial f_\alpha}{\partial r} = 0 \rightarrow f_\alpha\left[v_r(r), r\right] = f_{\alpha 0}\left[v_r(r_0), r_0\right] \qquad (4.14)$$

Here r_0 is the heliocentric distance to the exobase, v_r (and r) is the radial velocity (and the heliocentric distance) of a driving center of the solar wind α-component. The non-radial velocity component (in essence, the cyclotron rotation velocity) is determined from the condition that during the driving center motion along the magnetic field line, its transverse momentum component to magnetic field value ratio remains constant to within first-order infinitesimals, i.e. is the adiabatic invariant. In the solar gravitation field with the potential:

$$\varphi_g(r) = -\frac{GM_s}{r}$$

and in external radial magnetic and electric fields with the potential φ_E, the total energy of a driving center is conserved:

$$\frac{d}{dr}\left(\frac{m_\alpha v^2}{2} + m_\alpha \varphi_g + Ze\varphi_E\right) = 0$$

From where

$$\frac{m_\alpha v^2(r)}{2} - m_\alpha \frac{GM_s}{r} + Ze\varphi_E(r) =$$
$$\frac{m_\alpha v^2(r_0)}{2} - m_\alpha \frac{GM_s}{r_0} + Ze\varphi_E(r_0)$$

$$v(r_0) = \left\{v^2(r) + \frac{2GM_s}{r_0}\left(1 - \frac{r_0}{r}\right) - \frac{2Ze}{m_\alpha}\left[\varphi_E(r_0) - \varphi_E(r)\right]\right\}^{1/2}$$

(4.15)

According to Eq. (4.14),

$$f_\alpha[r, v(r)] = f_{\alpha 0}[r_0,$$
$$\sqrt{v^2(r) + \frac{2GM_s}{r_0}\left(1 - \frac{r_0}{r}\right) - \frac{2Ze}{m_\alpha}[\varphi_E(r_0) - \varphi_E(r)]}]$$

$Z = 1$ for protons and $Z = -1$ for electrons. Pikel'ner (Пикельнер 1950, 34-58) considered the motion above the exobase in the Pannekoek-Rosseland external electric field (see Eq. (4.12)). From Eq. (4.15), one obtains:

$$v^2(r) = v^2(r_0) - 2\frac{GM_s}{r_0}\left(1 - \frac{r_0}{r}\right)\left(1 - \frac{Z}{2}\frac{m_p}{m_\alpha}\right)$$

For protons ($Z = 1$, $m_\alpha = m_p$),

$$v_p^2(r) = v_p^2(r_0) - \frac{GM_s}{r_0}\left(1 - \frac{r_0}{r}\right)$$

For $r \to \infty$,

$$v_p^2(\infty) = v_p^2(r_0) - \frac{GM_s}{r_0} \tag{4.16}$$

$v_p^2(\infty) > 0$ if $v_p(r_0) > (GM_s/r_0)^{1/2}$.

For electrons ($Z = -1$, $m_\alpha = m_e$),

$$v_e^2(r) = v_e^2(r_0) - 2\frac{GM_s}{r_0}\left(1 - \frac{r_0}{r}\right)\left(1 + \frac{m_p}{2m_e}\right) \cong$$

$$v_e^2(r_0) - \frac{GM_s}{r_0}\left(1 - \frac{r_0}{r}\right)\frac{m_p}{m_e}$$

For $r \to \infty$,

$$v_e^2(\infty) = v_e^2(r_0) - \frac{GM_s}{r_0}\frac{m_p}{m_e} \tag{4.17}$$

$v_e^2(\infty) > 0$ if $v_p(r_0) > (m_p/m_e)^{1/2}(GM_s/r_0)^{1/2}$.

If the exobase is situated at the heliocentric distance $r_0 = (2–3)\ R$, then the conditions $v_p^2(\infty) > 0$ and $v_e^2(\infty) > 0$ are realizable as it is clear from Tables 4-1 and 4-2. If one specifies Maxwell's velocity distributions at the coronal base as follows:

$$f_{\alpha 0} = n\left(\frac{m_\alpha}{2\pi kT}\right)^{3/2} \exp\left(-\frac{m_\alpha v_0^2}{2kT}\right),$$

$$v_0^2 \equiv v^2(r_0) = \begin{cases} \dfrac{GM_s}{r_0}, \alpha = p - protons \\ \dfrac{GM_s}{r_0}\dfrac{m_p}{m_e}, \alpha = e - electrons \end{cases}$$

taking into account that $n_p(r_0) = n_e(r_0) = n$, $T_p(r_0) = T_e(r_0) = T$, then the solar wind proton velocity at the Earth orbit ($r = 215R$)

$$V(215R) = \frac{\int_{(GM_s/r_0)^{1/2}}^{\infty} v(215R) f_p[v(215R)]dv}{\int_{(GM_s/r_0)^{1/2}}^{\infty} f_p[v(215R)]dv}$$

proves to be several times lower than the measured on spacecraft. Because the free path is proportional to the fourth power of the particle velocity (see Eq. (4.3)), fast protons created below the exobase have a non-zero chance to escape into the interplanetary space. It ensures increasing the terminal solar wind proton velocity up to 140 km s^{-1} (Пикельнер 1950, 255-258). The comparison of Eq. (4.16) and Eq. (4.17) shows that the flux of escaping electrons is $(m_p/m_e)^{1/2}$ times higher than the flux of escaping protons. It means that there is the radial volume electric current in the solar wind, and it is incompatible with the magnetic field specified by Eq. (4.5). Hence, the Pannekoek-Rosseland potential describes the polarization electric field inadequately above the exobase. Jokers (Jokers 1970, 219-239) noted that at large heliocentric distances, the solar gravitation was negligible, the proton velocity – almost constant, the electron concentration decreased as r^{-2}, the electron temperature – as $r^{-1/2}$, the polarization electric field was balanced by the electron pressure gradient divided by the electron concentration and, hence, decreased as $r^{-3/2}$, i.e. slower than the solar gravitation. It means that the total (gravitation plus electric) potential has a maximum at some heliocentric distance above the exobase. Above this heliocentric distance, the electric force (accelerating protons) exceeds the gravitation force (confining protons), and this ensures increasing the calculated velocity of solar wind protons. Jokers specified Maxwell's velocity distributions for protons and electrons and equal fluxes of protons and electrons at the exobase (to ensure zero volume electric current everywhere in the solar wind). He determined the unknown electric potential heliocentric distance dependence by iterations equating calculated electron and proton concentrations (functions of the electric potential) at any heliocentric distance to ensure quasi-neutrality. According to his calculation, the solar wind velocity at the Earth orbit was 322 km s^{-1}, the concentration – 11 cm^{-3}, the proton temperature $T_p \sim 5.6\times10^4$ °K ($T_{p\|}/T_{p\perp} \sim 790$), and the electron temperature $T_e \sim 4.5\times10^5$ °K ($T_{e\|}/T_{e\perp} \sim 1$). When returning to Eq. (4.3), let us note that Eq. (4.3) permits the existence of non-Maxwell's velocity distributions sufficiently deep in the solar corona even in the transition region where $Kn \sim 10^3$ (Shoub 1983, 339-369), and this has observational confirmations (Esser and Edgar 2000, L71-L74). Maksimovich, with colleagues (Maksimovich et al. 1997, 725-734), specified non-Maxwell's velocity distributions (Kappa functions) at the exobase. The non-Maxwell's tail of supra-thermal electrons leads to the increase of the flux of escaping electrons. Hence, higher potential of the polarization electric field is needed to preserve the quasi neutrality of the solar wind. This enhanced electric field accelerates protons to terminal velocities up to 750 km s^{-1}.

The main idea of such kinetic constructions is to organize somehow the regime with large fluxes of accelerated supra-thermal ions and electrons from the Sun. Then the evaporation of scorching (and, hence, fully ionized) hydrogen plasma of the solar corona could be naturally explained by the difference between proton and electron masses under the equality of its electric charges.

Magneto-hydrodynamic description of the solar wind

Though the solar wind can be interpreted kinetically as a manifestation of the evaporation of hot coronal gases, mathematical constructions applied for the kinetic modelling of the phenomenon are too rigid: only the spherically symmetric solar wind in the radial magnetic field can be modeled. Magneto-hydrodynamics provides a more flexible mathematical instrument to model the spatially inhomogeneous magnetized solar wind in its acceleration region. Magneto-hydrodynamic equations interlink the first three moments of the velocity distribution. The first group of the equations is obtained via the multiplication of Eq. (4.4) by the electron (proton) mass with the subsequent integration over velocities. The second group of equations (in fact, three equations) is obtained via the multiplication of Eq. (4.4) by the corresponding component of an electron (a proton) momentum with the subsequent integration over velocities. The third group of equations is obtained via the multiplication of Eq. (4.4) by the kinetic energy of an electron (a proton) with the subsequent integration over velocities. Right-hand sides (with the participation of Boltzmann's integral) of all the equations prove to be zero: collisions change neither the total number of particles nor its total momentum and kinetic energy. The addition of the equations for electrons and protons of the first group gives the law of mass conservation as follows:

$$\frac{\partial \rho}{\partial t} + \nabla \cdot \rho \vec{V} = 0$$

Here

$$\rho(\vec{r},t) = n_e m_e + n_i m_i \tag{4.18}$$

$$\vec{V}(\vec{r},t) = \frac{m_e n_e \vec{V}_e + m_i n_i \vec{V}_i}{m_e n_e + m_i n_i} \tag{4.19}$$

The multiplication of the equations for electrons and protons of the first group by the elementary electric charge with the subsequent addition of the two obtained equations gives the electric charge conservation law as follows:

$$\frac{\partial \rho_q}{\partial t} + \nabla \cdot \vec{j} = 0$$

Here

$$\rho_q(\vec{r},t) = e(n_i - n_e) \tag{4.20}$$

$$\vec{j}(\vec{r},t) = e(n_i \vec{V}_i - n_e \vec{V}_e) \tag{4.21}$$

The addition of the two vector equations of the second group gives the law of momentum conservation as follows:

$$\rho \frac{\partial \vec{V}}{\partial t} + \rho\left(\vec{V} \cdot \nabla\right)\vec{V} = \rho_q \vec{E} + \frac{1}{c}\left[\vec{j} \times \vec{B}\right] - \nabla P - \rho \nabla \varphi_g \tag{4.22}$$

Here $P = n_e k T_e + n_i k T_i$. One may consider the pressure as a scalar when the Reynolds number is significant

$$\mathrm{Re} \equiv L V_0 \frac{m_i}{kT} \nu_{ii} >> 1$$

Here L is the character length, V_0 is the character speed, ν_{ii} is the ion-ion collision rate. The Reynolds number is high in the solar wind. The multiplication of the two vector equations of the second group by the electron charge to mass ratio and the proton charge to mass ratio, correspondingly, with the subsequent addition of obtained equations, gives the Ohm law as follows:

$$\vec{j} = \xi\left(\vec{E} + \frac{1}{c}\left[\vec{V} \times \vec{B}\right]\right) \tag{4.23}$$

Here $\xi = n_e e^2/(m_e \nu)$ is the electric conductivity, ν is the mean collision rate. The tremendous characteristic length of the solar wind phenomenon is taken into account when obtaining Eq. (4.23). The addition of the third group equations (taking into account Eq. (4.22) scalar multiplied by the velocity vector defined by Eq. (4.19)) under the assumption $T_i = T_e = T$ gives the energy conservation law as follows:

$$\frac{3k}{m_i}\rho\frac{dT}{dt} + \rho\left(\nabla \cdot \vec{V}\right) = Q \tag{4.24}$$

Here Q is the difference between emitted and absorbed energy of unit volume every second. Taking into account the condition $m_e/m_i \ll 1$ allows rewriting Eqs. (4.18) and (4.19) as follows:

$$\rho(\vec{r},t) \cong m_i n_i \tag{4.25}$$

$$\vec{V}(\vec{r},t) \cong \vec{V}_i \tag{4.26}$$

It means that the mass density of the solar wind is, in essence, the proton density and the solar wind velocity – the proton velocity. As the solar wind plasma is electrically neutral ($n_i = n_e = n$) on the spatial scale of the order of the Debye radius, one may rewrite Eqs. (4.20)-(4.22) to describe the solar wind on the spatial scale much larger than the Debye radius as follows:

$$\rho_q = 0 \tag{4.27}$$

$$\vec{j} = en\left(\vec{V}_i - \vec{V}_e\right) \tag{4.28}$$

$$m_i n \frac{\partial \vec{V}}{\partial t} + m_i n\left(\vec{V} \cdot \nabla\right)\vec{V} = \frac{1}{c}\left[\vec{j} \times \vec{B}\right] - \nabla P - m_i n_i \nabla \varphi_g \tag{4.29}$$

$$P = 2nkT \tag{4.30}$$

Let us note that the information about the polarization electric field of the solar wind is lost in Eqs. (4.24)-(4.30): magneto-hydrodynamics considers its influence implicitly via the thermal pressure gradient in the momentum equation.

The so-called single-fluid magneto-hydrodynamic equation set written above is probably the simplest mathematical construction that describes the self-consistent interaction between the plasma and the magnetic field.

This interaction leads to the nonstationary spatial inhomogeneous structure of the solar wind acceleration region.

Parker's model

Let us describe the pioneering among solar wind models – the Parker model published in 1958. Parker (Parker 1958, 644-675) considered the stationary hydrodynamic spherically symmetric solar wind without considering the force interaction between the plasma and the magnetic field. For simplicity, he assumed the polytrope relationship between the pressure and the density; however, the basic properties of his model could easier be illustrated for the isothermal corona, which is a particular case of the polytrope corona. One can eliminate the plasma concentration n from mass and momentum conservation equations:

$$\frac{1}{r^2}\frac{d}{dr}\left(r^2 nV\right)=0$$

$$m_p nV\frac{dV}{dr}=-2kT\frac{dn}{dr}-m_p n\frac{GM_s}{r^2}$$

to obtain the equation for the flow velocity as follows:

$$\frac{1}{V}\frac{dV}{dr}=\frac{\dfrac{4kT}{m_p r}-\dfrac{GM_s}{r^2}}{V^2-\dfrac{2kT}{m_p}} \qquad (4.31)$$

It is clear from Eq. (4.31) that the flow must pass the sonic barrier (where $V^2 = V_c^2 = 2kT/m_p$) at the heliocentric distance $r = r_c = GM_s m_p/(4kT)$: the other variants lead to the infinite flow acceleration (dV/dr) when passing the sonic barrier. Introducing the Mach number $M = V/V_c$, one rewrites Eq. (4.31) as follows:

$$\frac{1}{V}\frac{dV}{dr}\left(M^2 - 1\right) = 2\frac{d}{dr}\left(\ln r + \frac{r_c}{r}\right) \tag{4.32}$$

The function $f(r) = \ln r + r_c/r$ decreases ($df < 0$) up to the point $r = r_c$, where $df/dr = 0$, and increases ($df > 0$) after the passage of this point. Therefore, solutions of Eq. (4.32) with the continuously increasing ($dV > 0$) with heliocentric distance positive ($V > 0$) velocity are possible only when the flow is subsonic ($M < 1$) for $r < r_c$ and supersonic ($M > 1$) for $r > r_c$. The integration of Eq. (4.31) under the coronal temperature $T = 10^6$ °K gives the solar wind velocity at the Earth orbit slightly higher than 450 km s^{-1}.

Eq. (4.32) arises in hydrodynamics in calculations of one-dimensional gas flows in nozzles. The flow velocity is supposed to be directed along the nozzle axis while values of the velocity V, the pressure P, the density ρ, and the cross-section area S are assumed to depend on the distance x from the nozzle beginning only. The dependence $S(x)$ is assumed to be known. If one transforms the Euler equation

$$V\frac{dV}{dx} = -\frac{1}{\rho}\frac{dP}{dx}$$

to the form

$$VdV = -\frac{dP}{\rho} = -\frac{dP}{d\rho}\frac{d\rho}{\rho} = -V_{sound}^2\frac{d\rho}{\rho}$$

(V_{sound}^2 is the sonic speed square) and differentiates the continuity equation

$$\rho VS = const$$

to the form

$$\frac{d\rho}{\rho} = -\frac{dV}{V} - \frac{dS}{S}$$

substituting the latter expression into the right-hand side of the transformed Euler's equation, then one obtains the Hugoniot equation (Hugoniot 1886, 1178-1181):

$$\left(V^2 - V_{sound}^2\right)\frac{dV}{V} = V_{sound}^2\frac{dS}{S} \rightarrow \left(M^2 - 1\right)\frac{dV}{V} = \frac{dS}{S}$$

Here $M = V/V_{sound}$ is the Mach number. Looking to the Hugoniot equation, one concludes:

1. If $M < 1$, then the sign of dV is opposite to the sign of dS, i.e. the more significant is the cross-section area, and the slower is the subsonic flow.

2. If $M > 1$, then the signs of dV and dS are the same, i.e. the more significant is the cross-section area, and the faster is the supersonic flow. The density of supersonic flow in the divergent nozzle falls so deep that the product ρS decreases despite S increases; therefore, the flow velocity should increase to preserve the mass flow.
3. If $M = 1$, then $dS = 0$ so that the nozzle cross-section will be a critical (extremal) one. It is easy to conclude that the critical cross-section will be the cross-section of the minimal area because when approaching the maximal cross-section area, the subsonic flow decelerates while the supersonic one accelerates, and this cannot lead to the flow velocity equal to the sonic speed ($M = 1$) at the critical cross-section.
4. If $dS = 0$, then $dV = 0$ in the case of $M \neq 1$. In the subsonic regime, the flow velocity is minimal at the maximal cross-section area and maximal at the minimal cross-section area and vice versa in the supersonic regime.

In the practice of pressure calculations inside the nozzle, one specifies M_1, P_1, and S_1 (the Mach number, the pressure, and the cross-section area) at the given place of the nozzle and determines the pressure distribution of the isentropic flow along the nozzle axis parametrically as follows (Лойцянский, 1973, 130):

$$\frac{S}{S_1} = \frac{M_1}{M}\left(\frac{1+\dfrac{\eta-1}{2}M^2}{1+\dfrac{\eta-1}{2}M_1^2}\right)^{(\eta+1)/[2(\eta-1)]} \tag{4.33}$$

$$\frac{P}{P_1} = \left(\frac{1+\dfrac{\eta-1}{2}M_1^2}{1+\dfrac{\eta-1}{2}M^2}\right)^{\eta/(\eta-1)} \tag{4.34}$$

Here $\eta = C_p/C_v$, C_p is the heat capacity under constant pressure, C_v is the heat capacity under constant volume. Specifying the function $S(x)$, one calculates the function $M(x)$ from Eq. (4.33) at first and then calculates the function $P(x)$ from Eq. (4.34). Let us consider the isentropic flow from a reservoir of the huge volume to a chamber of the given backpressure through the Laval nozzle, which subsequently assembled a convergent tube (situated near a reservoir) and a divergent tube (situated near a chamber). In such a nozzle, the subsonic flow velocity increases in a convergent tube and, after the passage of the critical cross-section, can achieve supersonic velocities in a divergent tube in case of the right choice of the backpressure value. If one knows the minimal area of the nozzle S_1 and its position, then one puts $M_1 = 1$ there and calculates the pressure at the exit of the divergent tube by Eqs. (4.33)-(4.34) for $M > 1$. If this calculated pressure is equal to the backpressure, then the flow at the exit of the divergent tube is supersonic. Let us emphasize that there are many fully subsonic flows in the Laval nozzle of the given shape, but unique transonic flow exists and can be realized for the unique value of the backpressure. The supersonic velocity at the exit of Laval's nozzle can be as much increased (by adjusting the shape and the nozzle length) as lower the backpressure is set. One can imagine the Laval nozzle of the ideal shape to accelerate the flow to the supersonic velocity under zero backpressure.

The left-hand sides of Eq. (4.32) and the Hugoniot equation are the same. Comparing the right-hand sides of the equations, one concludes that unique transonic isothermal flow of the coronal material into the interplanetary space (the solar wind) could exist similar to the existence of unique transonic flow through the Laval nozzle of the given shape. Converting chaotic thermal motions to predominantly directed ones is the primary physical process in both cases. In the solar wind, the solar gravitation confines the hot plasma and creates the analog of the critical cross-section. At the same time, the backpressure is the very low pressure in the local interstellar medium. In the Laval nozzle, the conversion is ensured by shaped nozzle boundaries. Let us rewrite Eq. (4.31) as follows:

$$\frac{d}{dr}\left(\frac{V^2}{V_c^2} - \ln V^2\right) = 4\frac{d}{dr}\left(\ln r + \frac{r}{r_c}\right)$$

It is easy to see that only the velocity square is present in the above equation. It means that Eq. (4.31) is invariant concerning the flow direction, and besides the solar wind, the model describes the accretion of the interplanetary gas to the Sun. The mathematical tool to describe the accretion of the polytrope gas in the central gravitation field of a star had been constructed by Bondi (1952, 195-204) six years before Parker did.

Relationships between the solar wind and the solar rotation

Here we consider the influence of the solar rotation on the solar wind and vice versa. Electric currents induced in the solar wind because of the solar rotation are supposed to be the agents of the influence. The consideration is carried out in the non-inertial reference frame rotating with the Sun. The polar solar wind is considered separately from the near-equatorial one. In the end, we consider the solar wind influence on the solar rotation slowing down.

The Sun can be considered as a magnetized electrically conducting ball so that the unipolar induction will occur in case of the ball axial rotation (Ландау, Лифшиц, 1992, 327). The phenomenon is as follows: if the equator and the pole of the axially rotating ball are connected via a wire with sliding contacts, then the electric current will be registered in the wire. The electromotive force is easy to calculate in the non-inertial reference frame rotating with the ball. Suppose $\boldsymbol{\omega}$ is the angular velocity of the ball rotation. In the rotating frame, the wire rotates with the angular velocity $-\boldsymbol{\omega}$ (while the ball is fixed) so that everything is reduced to the motion of a conductor in the given magnetic field of the immoveable ball. The electromotive force acting between ends of a wire is given as the integral:

$$\mathrm{E} = \frac{1}{c}\int\left[\vec{V}\times\vec{B}\right]d\vec{l} = \frac{1}{c}\int\left[\vec{B}\times\left[\vec{r}\times\vec{\omega}\right]\right]d\vec{l}$$

taken along the wire. In the solar case, the role of an external wire belongs to the solar wind, where the solar rotation induces electric currents. The solar wind electric circuit contains the quasi azimuthal current flowing along spirals of Archimedes to the Sun in the near-equatorial solar wind, closed by radial currents in the polar solar wind directed, mainly, from the Sun. The rotating Sun acts as a unipolar inductor of the electromotive force (Alfven 1977, 271-284). Therefore, the magnetic field specified by Eq. (4.5) and requiring the absence of volume electric currents in the solar wind is incompatible with the solar rotation.

Let us choose the non-inertial reference frame rotating with the angular velocity $\boldsymbol{\omega}$ of the Sun. According to the field theory, the generalization of the Maxwell equations to the non-inertial reference frame is reduced to the construction of both covariant g_{nm} and contravariant g^{nm} metric tensor components. Then the contravariant field tensor:

$$F^{nm} = g^{nk} g^{mp} F_{kp}$$

allows writing the inhomogeneous Maxwell equations as follows:

$$\frac{1}{\sqrt{-g}} \frac{\partial}{\partial x^m} \left\{ \sqrt{-g} F^{nm} \right\} = \frac{4\pi}{c} J^n$$

(g is the determinant built on metric tensor components). The homogenous Maxwell equations have the same form in any (including non-inertial) frame of reference. Let us consider a slowly rotating frame of reference. It limits the spatial scale L to the value satisfying the condition $\omega L \ll c$ (c is the velocity of light). Due to the slow rotation, we take into account only linear terms relative to the ratio of characteristic velocity to velocity of light in the expression for the metric tensor (Grigor'ev et al., 2005, 62):

$$g_{nm} \cong \begin{pmatrix} 1 & \frac{\omega y}{c} & \frac{-\omega x}{c} & 0 \\ \frac{\omega y}{c} & -1 & 0 & 0 \\ \frac{-\omega x}{c} & 0 & -1 & 0 \\ 0 & 0 & 0 & -1 \end{pmatrix}$$

The axis z of the Cartesian coordinates is directed along the angular velocity vector. In the linear approximation, $g_{nm} = g^{nm}$, while gravitation amendments (in the Newton approximation) are minor and deviate beyond the framework of the interested accuracy. Covariant components of the field tensor

$$F_{nm} = \begin{pmatrix} 0 & -E_x & -E_y & -E_z \\ E_x & 0 & B_z & -B_y \\ E_y & -B_z & 0 & B_x \\ E_z & B_y & -B_x & 0 \end{pmatrix}$$

give the contravariant components as follows:

$$F^{nm} = g^{nk} g^{mp} F_{kp} = \begin{pmatrix} 0 & \widetilde{E}_x & \widetilde{E}_y & \widetilde{E}_z \\ -\widetilde{E}_x & 0 & \widetilde{B}_z & -\widetilde{B}_y \\ -\widetilde{E}_y & -\widetilde{B}_x & 0 & \widetilde{B}_x \\ -\widetilde{E}_z & \widetilde{B}_y & -\widetilde{B}_x & 0 \end{pmatrix} \tag{4.35}$$

Here $\widetilde{\vec{E}} \equiv \boldsymbol{E}\text{-}[\boldsymbol{\gamma}\times\boldsymbol{B}]$, $\widetilde{\vec{B}} \equiv \boldsymbol{B}\text{-}[\boldsymbol{\gamma}\times\boldsymbol{E}]$, $\boldsymbol{\gamma} = [\boldsymbol{\omega}\times\boldsymbol{r}]/c$, r is the heliocentric distance. Taking into account that in the linear approximation $(-g)^{1/2} \cong 1$, the inhomogeneous Maxwell equations are as follows:

$$\frac{\partial}{\partial x^m} F^{nm} = \frac{4\pi}{c} J^n \tag{4.36}$$

Substituting Eq. (4.35) to Eq. (4.36), one obtains:

$$\left[\nabla \times \widetilde{\vec{B}}\right] = \frac{1}{c}\frac{\partial}{\partial t}\widetilde{\vec{E}} + \frac{4\pi}{c}\vec{j}, \left(\nabla \cdot \widetilde{\vec{E}}\right) = 4\pi\rho_q \tag{4.37}$$

Suppose one applies the abstract function of the divergence operator to the first of Eqs. (4.37) and takes into account the second of Eqs. (4.37). In that case, one derives the electric charge conservation law as follows:

$$\frac{1}{c}\frac{\partial}{\partial t}\rho_q + \left(\nabla \cdot \vec{j}\right) = 0$$

The homogenous Maxwell equations have the same form in any (including non-inertial) frame of reference:

$$\left[\nabla \times \vec{E}\right] = -\frac{1}{c}\frac{\partial}{\partial t}\vec{B}, \left(\nabla \cdot \vec{B}\right) = 0$$

Let us transit to dimensionless variables choosing L^* as a length unit, V^* as a velocity unit, L^*/V^* as a time unit, B^* as a magnetic field unit, B^*V^*/c as an electric field unit, $B^*c/(4\pi L^*)$ as an electric current unit. Limiting to the linear relative to V^*/c approximation, we obtain:

$$\tilde{\vec{B}} = \vec{B}, \tilde{\vec{E}} = \vec{E} - \frac{1}{Ro}\left[\left[\vec{k} \times \vec{r}\right] \times \vec{B}\right] \tag{4.38}$$

($Ro = V^*/(\omega L^*)$ is the Rossby number), and the Maxwell equations are as follows:

$$\left[\nabla \times \vec{B}\right] = \vec{j}, \left(\nabla \cdot \tilde{\vec{E}}\right) = 4\pi\rho_q, \left[\nabla \times \vec{E}\right] = -\frac{\partial}{\partial t}\vec{B}, \left(\nabla \cdot \vec{B}\right) = 0$$

The Ohm law in dimensionless variables is as follows:

$$\vec{j} = \mathrm{Re}_m\left(\vec{E} + \left[\vec{V} \times \vec{B}\right]\right)$$

Here $Re_m = 4\pi\xi L^* V^*/c^2$ is the magnetic Reynolds number (ξ is the electric conductivity). Suppose one does not consider small scale processes in the vicinity of interplanetary magnetic field neutral surfaces and inside double electric layers or interplanetary shocks. In that case, high character velocities (hundreds of km s^{-1}), long character lengths (of the order of the distance between the Sun and the Earth) along with the relatively good electric conductivity of the solar wind plasma (which is comparable with the electric conductivity of bad metallic conductors) lead to very high values of the magnetic Reynolds number. In the approximation $Re_m \to \infty$, the Ohm law degenerates as follows:

$$\vec{E} = -\left[\vec{V} \times \vec{B}\right]$$

Let us consider the polar solar wind. The joint analyses of the "Ulysses" spacecraft radio signal propagation through the polar solar wind to the ground station and observations of polar K- and X-ray coronae allows tracing electron concentration inhomogeneities in the heliocentric distance interval (1.03–30) R. It was found that polar coronal hole boundaries, as well as bright coronal points, were projected to the polar solar wind almost radially (Woo and Habbal 1999, L69-L72). It permits us to suppose the magnetic field specified by Eq. (4.5) as the magnetic field of the polar solar wind in a zero-order approximation. The only field-aligned electric currents induced by the solar rotation are most probably present in the polar solar wind (Alfven 1977, 271-284); therefore, the polar solar wind magnetic field can be assumed force-free. Let us choose R as the unit of length, the solar escape velocity $(2GM_s/R)^{1/2}$ – as the unit of velocity, the coronal base density ρ_s – as the unit of density, $2GM_s\rho_s/R$ – as the unit of pressure, B_s – as the unit of the magnetic field, $(B_s/c)(2GM_s/R)^{1/2}$ – as the unit of the electric field, $B_sc/(4\pi R)$ – as the unit of electric current density, $c/(4\pi R)$ – as the unit of twisting parameter.

Stationary dimensionless magneto-hydrodynamic equations in the rotating frame of reference are as follows:

$$\left[\nabla \times \vec{E}\right] = 0 \rightarrow \vec{E} = -\nabla \varphi$$

$$(\nabla \cdot \nabla \varphi) + \frac{1}{Ro} r\left(\nabla \cdot \left[\left[\vec{k} \times \vec{e}_r\right] \times \vec{B}\right]\right) = 0$$

$$-\nabla \varphi + \left[\vec{V} \times \vec{B}\right] = 0, (V \cdot \nabla \varphi) = 0$$

$$\left[\nabla \times \vec{B}\right] = \alpha \vec{B}, \left(\nabla \cdot \vec{j}\right) = 0 \rightarrow \left(\vec{B} \cdot \nabla \alpha\right) = 0$$

$$\left(\nabla \cdot \vec{B}\right) = 0$$

$$\left(\nabla \cdot \rho \vec{V}\right) = 0$$

$$\rho\left(\vec{V} \cdot \nabla\right)\vec{V} = -\nabla P - \frac{\rho}{r^2}\vec{e}_r + \frac{2}{Ro}\rho\left[\vec{V} \times \vec{k}\right] - \frac{1}{Ro^2}\rho r\left[\vec{k} \times \left[\vec{k} \times \vec{e}_r\right]\right] \qquad (4.39)$$

Here $\boldsymbol{k} = \boldsymbol{\omega}/\omega$, $\boldsymbol{e}_r = \boldsymbol{r}/r$, $Ro = (2GM_s/R)^{1/2}/(\omega R)$ is the Rossby number, α is the twisting parameter. The density of the volume electric charge is assumed zero; this implies that the characteristic spatial scale should be much larger than the Debye radius and the difference between ion and electron velocities (characterizing the electric current density) – much smaller than the solar wind velocity. Let us consider the lateral boundary of the polar solar wind to be the lateral boundary of the cone with the axis coinciding with the axis of the solar rotation, the top – the center of the Sun, θ_b – the top-angle (parameter). The inverted Rossby number Ro^{-1} is ~ 10^{-3} for the sidereal period of the polar coronal base rotation of about 30 days and the solar escape velocity equal to 617.7 km s^{-1}. Thus the value Ro^{-1} can be used as a small parameter, and we can search the solution to Eqs. (4.39) as follows:

$$\varphi = \varphi_0 + \frac{1}{Ro}\varphi_1, \alpha = \alpha_0 + \frac{1}{Ro}\alpha_1, \rho = \rho_0 + \frac{1}{Ro}\rho_1, P = P_0 + \frac{1}{Ro}P_1$$

$$\vec{V} = \vec{V}_0 + \frac{1}{Ro}\vec{V}_1, \vec{B} = \vec{B}_0 + \frac{1}{Ro}\vec{B}_1 \qquad (4.40)$$

Substituting Eqs. (4.40) into Eqs. (4.39) and neglecting the small terms O(Ro^{-2}), we obtain two systems – the system of zero-order approximation (the absence of rotation) and the system of first-order approximation (the slow rotation). The system of zero-order approximation (the absence of rotation) is as follows:

$$(\nabla \cdot \nabla \varphi_0) = 0$$

$$-\nabla \varphi_0 + [\vec{V}_0 \times \vec{B}_0] = 0$$

$$(\vec{V}_0 \cdot \nabla \varphi_0) = 0$$

$$[\nabla \times \vec{B}_0] = \alpha_0 \vec{B}_0$$

$$\left(\vec{B}_0 \cdot \nabla \alpha_0\right) = 0$$

$$\left(\nabla \cdot \vec{B}_0\right) = 0$$

$$\left(\nabla \cdot \rho_0 \vec{V}_0\right) = 0$$

$$\rho_0 \left(\vec{V}_0 \cdot \nabla\right)\vec{V}_0 = -\nabla P_0 - \frac{\rho_0}{r^2}\vec{e}_r \tag{4.41}$$

Let us choose the magnetic field specified by Eq. (4.5) and the Parker isothermal (T is the solar wind temperature) spherically symmetric solution as a zero-order approximation for the polar solar wind. In the spherical coordinate system (r, θ, $\varnothing$) ($\varnothing$ is the heliolongitude in the rotating frame of reference), it is as follows:

$$\varphi_0 = const, \alpha_0 = 0, \vec{B}_0 = \left\{\frac{B_s}{r^2}, 0, 0\right\}, \vec{V}_0 = \{V_{r0}(r), 0, 0\},$$

$$\rho_0 V_{r0} r^2 = \rho_s V_s, P_0 = C_4 \rho_0, C_4 = \frac{kTR}{GM_s m_p},$$

$$\frac{V_{r0}^2}{2} + C_4 \ln \frac{V_s}{V_{r0} r^2} + \frac{1}{2}\left(1 - \frac{1}{r}\right) = 0{,}3 \ln C_4 + \frac{1}{C_4} = -2\ln(16 V_s) \tag{4.42}$$

Here ρ_s, V_s, B_s are the density, the velocity, and the magnetic field at the coronal base $r = R$. The last of Eqs. (4.42) is the condition of the Parker integral curve passage through the critical point under the assumption V_s

$\ll 1$. For the chosen zero-order approximation, the system of first-order approximation (the slow rotation) is as follows:

$$
\begin{aligned}
&\left(\nabla \cdot \nabla \varphi_1\right) + \frac{B_s}{r}\left(\nabla \cdot \left[\left[\vec{k} \times \vec{e}_r\right] \times \vec{e}_r\right]\right) = 0 \\
&-\nabla \varphi_1 + \frac{B_s}{r^2}\left[\vec{V}_1 \times \vec{e}_r\right] + V_{ro}\left[\vec{e}_r \times \vec{B}_1\right] = 0 \\
&\left(\vec{e}_r \cdot \nabla \varphi_1\right) = 0 \\
&\left[\nabla \times \vec{B}_1\right] = \alpha_1 \frac{B_s}{r^2}\vec{e}_r \\
&\left(\vec{e}_r \cdot \nabla \alpha_1\right) = 0 \\
&\left(\nabla \cdot \vec{B}_1\right) = 0 \\
&\left(\nabla \cdot \rho_1 \vec{V}_0\right) + \left(\nabla \cdot \rho_0 \vec{V}_1\right) = 0 \\
&\rho_1\left(\vec{V}_0 \cdot \nabla\right)\vec{V}_0 + \rho_0\left(\vec{V}_1 \cdot \nabla\right)\vec{V}_0 + \rho_0\left(\vec{V}_0 \cdot \nabla\right)\vec{V}_1 = \\
&-\nabla P_1 - \frac{\rho}{2r^2} + 2\rho_0 V_{r0}\left[\vec{e}_r \times k\right]
\end{aligned}
\tag{4.43}
$$

The linear character of Eqs. (4.43) permits constructing its solution $\{\varphi_1, \alpha_1, \rho_1, P_1, \boldsymbol{V}_1, \boldsymbol{B}_1\}$ by the variable separation method (Pisanko 1997, 345-352); the details can be found in (Grigor'ev et al., 2005, 74-90). The result in the attendant inertial frame of reference is as follows:

$$P = C_4 \frac{\rho_s V_s}{V_{r0} r^2} + \frac{1}{Ro} \frac{\rho_s V_s V_{r0}}{r} \left[2 + \sqrt{Q(r)}\right] F_2(\theta, \varnothing) H(r - r_d) \quad (4.44)$$

$$\rho = \frac{\rho_s V_s}{V_{r0}^2 r^2} \left[V_{r0} - \frac{1}{Ro} F_1(r) F_2(\theta, \varnothing) H(r - r_d) \right] \quad (4.45)$$

$$V_r = V_{r0} + \frac{1}{Ro} F_1(r) F_2(\theta, \varnothing) H(r - r_d) \quad (4.46)$$

$$F_1(r) = \int_1^r V_{r0}(\xi) \left\{ 2 - Q(\xi) - 2\left[C_4 - V_{r0}^2(\xi) Q(\xi)\right] \cdot \left[V_{r0}^2(\xi) - C_4\right] - 1 \right\}$$

$$\exp\left[\int_\xi^r \frac{d\eta}{2\eta^2 V_{r0}^2(\eta)} \right] d\xi \quad (4.47)$$

$$Q(\xi) = \left[\frac{2C_4 - 0.5}{\xi V_{r0}^2(\xi) - C_4 \xi} \right]^2$$

$$F_2(\theta, \varnothing) = a_0 + \sum_{m=3}^{\infty} tg^m(\theta/2)(a_m \sin m\varnothing + b_m \cos m\varnothing)$$

$$V_\theta = \frac{1}{Ro} r V_{r0} F_3(\theta, \varnothing) H(r - r_d) \quad (4.48)$$

$$F_3(\theta, \varnothing) = \sum_{m=3}^{\infty} \frac{m \sin^{m-1}(\theta/2)}{2\cos^{m+1}(\theta/2)} (a_m \cos m\varnothing + b_m \sin m\varnothing)$$

$$V_\varnothing = \frac{1}{Ro} \left\{ r \sin\theta \left[1 - H(r - r_d)\right] + r V_{r0} F_4(\theta, \varnothing) H(r - r_d) \right\} \quad (4.49)$$

$$F_4(\theta,\varnothing) = \sum_{m=3}^{\infty} m tg^m(\theta/2)\left(a_m \sin m\varnothing - b_m \cos m\varnothing\right)$$

$$\varphi = C_5 - \frac{1}{Ro} B_s \cos\theta\left[1 - H(r - r_d)\right] \tag{4.50}$$

$$B_r = \frac{B_s}{r^2} \tag{4.51}$$

$$B_\theta = \frac{1}{Ro}\frac{B_s}{r} F_3(\theta,\varnothing) H(r - r_d) \tag{4.52}$$

$$B_\varnothing = \frac{1}{Ro}\frac{B_s}{r} F_4(\theta,\varnothing) H(r - r_d) \tag{4.53}$$

$$j_r = \frac{1}{Ro}\frac{B_s}{r^2} F_5(\theta,\varnothing) \tag{4.54}$$

$$F_5(\theta,\varnothing) = \sum_{m=3}^{\infty} \frac{m^2 \sin^{m-2}(\theta/2)}{2\cos^{m+2}(\theta/2)}\left(a_m \cos m\varnothing - b_m \sin m\varnothing\right)$$

Note that (in the attendant inertial frame of reference) $\varnothing$ denotes the a-dimensional combination $\varnothing_1$-ωt (where $\varnothing_1$ is the heliolongitude in the inertial frame of reference, t is time) so that Eqs. (4.44)-(4.54) will demonstrate the nonstationary (wavelike) heliolongitude dependence in the inertial frame of reference. Eqs. (4.44)-(4.54) are written formally for $\theta < \pi/2$; for $\theta > \pi/2$, the symmetry to the equatorial plane can be used. Here $H(r-r_d)$ is the Heaviside function (r_d is a parameter). The addendum ($m = 1$) is eliminated because of the $|j_r| < \infty$ and the addendum ($m = 2$) –

because of the j_r continuity at the polar axis. The value V_s is determined from the last of Eqs. (4.42) – the condition of the passage of V_{r0} through the critical point. Eq. (4.49) and Eq. (4.50) indicate that the polar corona above r_d does not co-rotate with the polar photosphere. It could be interpreted as the result of the coronal plasma zonal drift (above r_d) in crossed radial magnetic and meridian electric fields (analogous to the "homopolar effect" known in the nuclear fusion research). The drift angular velocity is equal to the angular velocity of the polar photosphere, but its direction is opposite. The result is the absence of the polar corona co-rotation above r_d. If the polar corona below r_d co-rotates with the polar photosphere and, hence, is under the potential of the polarization electric field, then an electric double layer (Alfven 1977, 271-284) exists at $r = r_d$. The double layer (with the non-zero radial electric field inside and the finite non-zero anomalous electric conductivity) produces the field-aligned volume electric current system.

The requirement of the absence of the surface-bound electric charge at the lateral boundary ($\theta = \theta_b$) is stated when constructing Eqs. (4.44)-(4.54), which means the continuity of the only non-zero θ-component of the electric field across the boundary. The latter can be ensured when the equatorial corona (for $r > r_d$) rotates with the angular velocity $\omega_1 > \omega$, and the magnetic field at the "equatorial side" of the lateral boundary has the helical configuration (Grigor'ev et al., 2005, 91)

$$B_r = \frac{B_s}{r^2}, B_{\varnothing} = \frac{\omega_1}{\omega}\frac{1}{Ro}\frac{B_s}{V_{r0}r}\sin\theta_b \qquad (4.55)$$

inclined to the west (i.e. in the direction of the solar rotation). It is interesting to note that sixteen years (1977-1992) of the Wilcox Solar Observatory magnetic data were used to determine the average east-west inclination of photospheric magnetic field lines, and accurate calculations showed the net tilt of 0.6° to the west (Scrauner and Scherrer 1994, 131-141). Eq. (4.53) and Eq. (4.55) interpret the polar solar wind lateral boundary above r_d as the sheet carrying the radial surface current:

$$i_r = \frac{1}{Ro}\frac{B_s}{r}\left[\frac{\omega_1}{\omega}\frac{1}{V_{r0}}\sin\theta_b - F_4(\theta,\varnothing)\right] \qquad (4.56)$$

The electric current system (described by Eq. (4.54) and Eq. (4.56)) differs from the Alfven concept (Alfven 1977, 271-284) of the solar rotation axis as a "wire" with an electric current. Instead, the polar corona can be treated as a "coaxial cable" – the current conical sheet (an envelope) with the surface current i_r flowing along the polar corona boundary ($\theta = \theta_b$) and a lot of "wires" – volume radial electric currents of opposite each other polarities inside the envelope. The average conical current sheet strength should be equal (and opposite) to the average near-equatorial interplanetary current sheet strength. The average over a solar polar cap volume electric current is equal to zero. The volume electric current density is equal to zero at the solar rotation axis and increases toward the lateral boundary of the polar corona. Supra-thermal tails of both ion and electron velocity distributions were registered in high heliolatitudes and near the equator (Gloeckler et al. 2000, 221-228) (Gloeckler 2003, 583-588) (Decker et al. 2006, 73-78) (Simunac and Armstrong 2004, https://doi.org/10.1029/2003JA010194). They could be considered the carriers of field-aligned currents. The minimum permissible number of "wires" is equal to six ($m = 3$). It is interesting to notice that the similar artificial current system (two coaxial coils and six wires carrying electric currents of opposite polarities parallel to the axis of coils) was constructed for creating the magnetic field in the nuclear fusion experimental installation at Kurchatov Institute of Atomic Energy (Арцимович, 1978, 92). These currents create transverse magnetic variations (similar to that observed in the Earth polar magnetosphere (Potemra 1986, 397-402)) and the "convective cell pattern" for non-radial components of the flow velocity via "frozen-in" conditions. In-situ magnetic observations on the "Ulysses" spacecraft revealed that the almost uniform radial magnetic field existed in the polar solar wind, while non-radial magnetic field components and flow velocity ones varied in-phase (Balogh et al. 1995,

1007-1010). It corresponds to Eqs. (4.48)-(4.49) and Eqs. (4.51)-(4.53). The additional pressure gradient (proportional to a_0) created by the field-aligned electric currents (supra-thermal ions and electrons) leads to the additional acceleration of the polar solar wind. Averaging Eqs. (4.44)-(4.46) over both heliolatitudes and heliolongitudes, one obtains:

$$\langle P \rangle = C_4 \frac{\rho_s V_s}{V_{r0} r^2} + a_0 \frac{1}{Ro} \frac{\rho_s V_s V_{r0}}{r} \left[2 + \sqrt{Q(r)}\right] H(r - r_d) \quad (4.57)$$

$$\langle \rho \rangle = \frac{\rho_s V_s}{V_{r0}^2 r^2} \left[V_{r0} - a_0 \frac{1}{Ro} F_1(r) H(r - r_d) \right] \quad (4.58)$$

$$\langle V_r \rangle = V_{r0} + a_0 \frac{1}{Ro} F_1(r) H(r - r_d) \quad (4.59)$$

Eqs. (4.44)-(4.59) are valid until terms proportional to Ro^{-2} can be considered small ones. It limits both the region of the definition of Eqs. (4.44)-(4.59) and variation intervals of the parameters a_0 and r_d. The parameters a_0 and r_d should be chosen, considering $Ro^{-1} V_1$ can be no more than 25–30% of the V. Otherwise, terms proportional to Ro^{-2} cannot be neglected. At the same time, values of the parameters a_0 and r_d should be no smaller than O(Ro^{-2}) to preserve consistency. When r_d is situated very close to the solar surface, the function $F_1(r)$ defined for $r > r_d$ has enormous values (see Eq (4.47)). Moving r_d away from the Sun, the values of $F_1(r)$ can be reduced. Numerical tests were done to determine optimal values of the parameters a_0, r_d, T satisfying the restrictions indicated above. The value of the parameter a_0 was obtained by the normalization of Eq (4.58) to a given empirical heliocentric distance dependence of electron concentration in the polar corona. The empirical dependence (Munro and Jackson 1977, 874-886) gives $a_0 = 0.632$, $r_d = 2.6\ R$, $T = 2\times10^6$ °K (Pisanko 1997, 345-352) (Pisanko 1997, 605-608), while the empirical dependence (Kohl et al. 1998, L127-L131) – $a_0 = 0.627$, $r_d = 2.4\ R$, $T = 2\times10^6$ °K (Pisanko 1998, 297-300). It gives $<V_r(30R)> \cong 750$ km s^{-1}, which is close to velocities observed by the "Ulysses" spacecraft in the polar solar wind in the solar minimum. Signs <…> mean averaging both heliolatitudes and heliolongitudes inside the cone ($\theta = \theta_b$). Eqs. (4.44)-(4.59) are limited from above at the heliocentric distance of (50–60) R. Otherwise, terms proportional to Ro^{-2} cannot already be neglected. It is interesting to notice that the value of $r_d = (2.4 - 2.6)\ R$ (the radial position of the "hypothetical" electric double layer) proves to be very close to the radial coordinate $r = 2.5\ R$ of the so-called "source surface" which was obtained by adapting the current-free magnetic field model to the white light eclipse image of the solar corona (Altschuler and Newkirk 1969, 131-149).

An alternative, not connected to the solar rotation, mechanism of the polar solar wind additional acceleration implies the additional pressure of

Alfven waves. The waves propagate from the solar convective zone to the polar solar wind and are effective when the radial magnetic field component at the base of the polar corona is no less than 6–7 Gauss (Чашей 1992, 192-200). However, estimates of the radial magnetic field component at the polar photosphere are very uncertain: magnetograph measurements permit to register only the photospheric magnetic field component directed along the line of sight, but when looking to solar polar caps, one rather measures the horizontal component than the radial one (the projection angle is too large).

Let us consider now the near-equatorial solar wind at heliocentric distances further than 30 *R*. Direct measurements on spacecraft show that the near-equatorial solar wind between orbits of Mercury and Saturn is supersonic superalfvenic and radially directed. Also, radio-sounding testifies for the supersonic superalfvenic wind already at heliocentric distances of about 20 *R* (Yakovlev and Pisanko 2018, 552-566). At heliocentric distances further than 30 *R*, one can neglect the solar gravitation compared with the inertia force. At the same time, the supersonic and superalfvenic flow character permits the neglect of the Ampere force and the pressure gradient. In such conditions, magneto-hydrodynamic equations degenerate as follows:

$$\left[\nabla \times \vec{B}\right] = \vec{j}$$

$$\left(\nabla \cdot \tilde{\vec{\vec{E}}}\right) = 4\pi\rho_q$$

$$\left[\nabla \times \left[\vec{V} \times \vec{B}\right]\right] = \frac{\partial}{\partial t}\vec{B}$$

$$\left(\nabla \cdot \vec{B}\right) = 0$$

$$\frac{\partial}{\partial t}\rho + \left(\nabla \cdot \rho\vec{V}\right) = 0$$

$$\frac{\partial}{\partial t}\vec{V} + \left(\vec{V} \cdot \nabla\right)\vec{V} = \frac{2}{Ro}\left[\vec{V} \times \vec{k}\right] - \frac{r}{Ro^2}\left[\vec{k} \times \left[\vec{k} \times \vec{e}_r\right]\right] \qquad (4.60)$$

Here $Ro = V/\omega L$ (V is the solar wind velocity at the Earth orbit, L is the distance between the Sun and the Earth). In spherical coordinates (r, θ, $\varnothing$) rotating with the Sun (the colatitude $\theta = 0$ at the solar north pole), the solar wind velocity components are specified as follows:

$$V_r = V_r\left(r, \theta, \varnothing, t\right), V_\varnothing = -\frac{1}{Ro} r \sin\theta, V_\theta = 0 \qquad (4.61)$$

The non-radial velocity component $V_\varnothing$ is completely stipulated by the choice of the rotating reference frame. By substituting Eqs. (4.61) into the last two of Eqs. (4.60), one obtains:

$$\frac{\partial}{\partial t}V_r + V_r\frac{\partial}{\partial r}V_r - \frac{1}{Ro}\frac{\partial}{\partial \varnothing}V_r = 0$$

$$\frac{\partial}{\partial t}\left(r^2\rho\right) + r^2\rho\frac{\partial}{\partial r}V_r + V_r\frac{\partial}{\partial r}\left(r^2\rho\right) = 0$$

Substituting Eqs. (4.61) into the third of Eqs. (4.60) and taking into account the condition:

$$\left(\nabla\cdot\vec{B}\right)=0\rightarrow\frac{1}{r}\frac{\partial}{\partial r}\left(r^2B_r\right)+\frac{\partial B_\theta}{\partial\theta}=-\frac{1}{\sin\theta}\left(B_\theta\cos\theta+\frac{\partial B_\varnothing}{\partial\varnothing}\right),$$

one obtains:

$$\frac{\partial}{\partial t}\left(rB_\theta\right)+V_r\frac{\partial}{\partial r}\left(rB_\theta\right)-\frac{1}{Ro}\frac{\partial}{\partial\varnothing}\left(rB_\theta\right)=-rB_\theta\frac{\partial V_r}{\partial r}$$

$$\frac{\partial}{\partial t}\left(rB_\varnothing\right)+V_r\frac{\partial}{\partial r}\left(rB_\varnothing\right)-\frac{1}{Ro}\frac{\partial}{\partial\varnothing}\left(rB_\varnothing\right)=\frac{1}{Ro}rB_\theta\cos\theta-rB_\varnothing\frac{\partial V_r}{\partial r}$$

$$\frac{\partial}{\partial t}\left(r^2B_r\sin\theta\right)+V_r\frac{\partial}{\partial r}\left(r^2B_r\sin\theta\right)-\frac{1}{Ro}\frac{\partial}{\partial\varnothing}\left(r^2B_r\sin\theta\right)=$$
$$rB_\theta\frac{\partial}{\partial\varnothing}V_r+rB_\varnothing\frac{\partial}{\partial\varnothing}V_r$$

Introducing characteristic equations:

$$\frac{dr}{dt}=V_r,\frac{d\varnothing}{dt}=-\frac{1}{Ro},\frac{d\theta}{dt}=0, \tag{4.62}$$

one obtains the equation for the velocity on a characteristic as follows:

$$\frac{d}{dt}V_r=0\rightarrow V_r=const \tag{4.63}$$

The equation for calculating B_θ along a characteristic is as follows:

$$\frac{d}{dt}\left(rB_\theta\right) = -rB_\theta \frac{\partial}{\partial r}V_r \rightarrow$$
$$rB_\theta = C_1 \exp\left\{-\int \frac{\partial V_r}{\partial r}dt\right\}, C_1 = const \tag{4.64}$$

The equation for calculating $B_\varnothing$ along a characteristic is as follows:

$$\frac{d}{dt}\left(rB_\varnothing\right) = -rB_\varnothing \frac{\partial}{\partial r}V_r + \frac{1}{Ro}rB_\theta \cos\theta \tag{4.65}$$

The equation for calculating B_r along a characteristic is as follows:

$$\frac{d}{dt}\left(r^2 B_r \sin\theta\right) = rB_\theta \sin\theta \frac{\partial V_r}{\partial \theta} + rB_\varnothing \frac{\partial V_r}{\partial \varnothing} \tag{4.66}$$

The equation for the solar wind density is as follows:

$$\frac{d}{dt}\left(r^2 \rho\right) = -r^2 \rho \frac{\partial V_r}{\partial r} \rightarrow$$
$$r^2 \rho = C_2 \exp\left\{-\int \frac{\partial V_r}{\partial r}dt\right\}, C_2 = const \tag{4.67}$$

In essence, Eq. (4.67) describes the solar wind inertial motion when the occurrence of corrugations (regions of the infinite density) increases with the heliocentric distance increase: fast particles overtake slow ones when Ampere's forces and thermal pressure gradients (that could prevent corrugations) are neglected. The nonlinear scheme (the demand for inhomogeneities to be small is absent in the scheme) (4.62)-(4.67) was proposed in (Шабанский и Шистер 1973, 825-828) (Шабанский и

Шистер 1973, 569-573). For known ***V*** and ***B***, the electric charge density in the solar wind is determined by the equation:

$$\left(\nabla \cdot \tilde{\vec{E}}\right) = 4\pi\rho_q$$

Here $\tilde{\vec{E}}$ is determined from the second of Eqs. (4.39) with known ***E*** = -[***V***×***B***] and ***B***. The following equation determines the electric current density ***j***:

$$\left[\nabla \times \vec{B}\right] = \vec{j}$$

If $V_r = const$, $V_\varnothing = -r\sin\theta / Ro$, $B_\theta = 0$, then Eqs. (4.62)-(4.67) give $rB_\varnothing = const$ and $r^2 B_r \sin\theta = const$ on a characteristic so that:

$$\frac{V_\varnothing}{V_r} = C_3 r \sin\theta = \frac{B_\varnothing}{B_r}, C_3 = const \qquad (4.68)$$

Eq. (4.68) means that the solar wind velocity is parallel to the magnetic field in the rotating reference frame. In the inertial frame, Eq. (4.68) is as follows:

$$B_r(r,\theta,\varnothing) = B_r\left[30R,\theta,\varnothing - \frac{\omega}{V_r}(r-30R)\right]\left(\frac{30R}{r}\right)^2$$
$$B_\theta(r,\theta,\varnothing) = 0 \qquad (4.69)$$
$$B_\varnothing(r,\theta,\varnothing) = B_r\left[30R,\theta,\varnothing - \frac{\omega}{V_r}(r-30R)\right]\frac{900R^2\omega\sin\theta}{V_r r}$$

All field lines of this magnetic field are open-ended spirals of Archimedes curled oppositely to the solar rotation. Parker (Parker 1958, 644-675) derived this result from intuitive arguments but Eqs. (4.63)-(4.67) are more powerful and include Parker's model as a particular case. For example, from Eq. (4.64), one can calculate the evolution of the interplanetary magnetic field north-south component (Parker's model cannot do this); while the comparison of Eq. (4.64) with Eq. (4.67) suggests an idea of the sufficiently high correlation between the magnetic field north-south component and the solar wind density. The solenoidal character of the field (4.69) is ensured by postulating the current equatorial sheet, which separates oppositely directed field lines above and below the equator. The surface current also flows along spirals of Archimedes but curled in the direction of the solar rotation. The current is almost azimuthal at small heliocentric distances but becomes almost radial at large heliocentric distances.

Let us consider the solar wind influence on the solar rotation slowing down in the stationary case of $\{V_r(r), 0, V_\varnothing(r)\}$ $\{B_r(r), 0, B_\varnothing(r)\}$ at the equatorial plane (Weber and Davis 1967, 217-227). The solenoidal character of the magnetic field requires:

$$B_r(r)r^2 = const \qquad (4.70)$$

If the electric charge density is zero, then the equations:

$$\left[\nabla \times \vec{E}\right] = 0$$

$$\left(\nabla \cdot \vec{E}\right) + \frac{1}{Ro} r\left(\nabla \cdot \left[\left[\vec{k} \times \vec{e}_r\right] \times \vec{B}\right]\right) = 0$$

degenerate at the equatorial plane as follows:

$$\left[\nabla \times \vec{E}\right] = 0, \left(\nabla \cdot \vec{E}\right) = 0 \qquad (4.71)$$

If ***V***||***B*** at the equator of the coronal base, then ***E*** = 0 (because ***E*** = -[***V***×***B***]). It means that ***E*** = 0 everywhere at the equatorial plane due to Eqs. (4.71), hence, ***V***||***B*** everywhere at the equatorial plane in the rotating reference frame. In the inertial frame of reference, the latter is written as follows:

$$\frac{B_r}{B_\varnothing} = \frac{V_r}{V_\varnothing - \omega r} \qquad (4.72)$$

Eq. (4.72) considers that the magnetic field is the same (in linear approximation relative to V/c) in rotating and inertial frames of reference. The projection of momentum to the azimuthal axis in the inertial frame of reference is as follows:

$$\rho \frac{V_r}{r} \frac{d}{dr}\left(r V_\varnothing\right) = \frac{B_r}{4\pi r} \frac{d}{dr}\left(r B_\varnothing\right) \qquad (4.73)$$

Taking into account Eq. (4.70) and the mass flow conservation

$$\rho V_r r^2 = const,$$

one can directly integrate Eq. (4.73) as follows:

$$rV_{\varnothing} - \frac{B_r}{4\pi\rho V_r} rB_{\varnothing} = Y = const \qquad (4.74)$$

Here Y is the constant angular moment per mass unit of the equatorial solar wind. Eq. (4.74) means that the angular moment taken away with the equatorial solar wind is transmitted to the base of the equatorial corona via the Ampere force, and it should slow down the solar rotation. If one defines the Alfven-Mach number as follows:

$$M_A^2 = \frac{4\pi\rho V_r}{B_r^2} \qquad (4.75)$$

and eliminates $B_{\varnothing}$ from Eq. (4.74) with the help of Eq. (4.75), one obtains:

$$V_{\varnothing} = \omega r \frac{M_A^2 Y \dfrac{1}{\omega r^2} - 1}{M_A^2 - 1}.$$

At the heliocentric distance where $M_A = 1$, the velocity component $V_{\varnothing}$ remains limited only in case:

$$\frac{Y}{\omega r_A^2} - 1 = 0 \rightarrow Y = \omega r_A^2$$

Hence, it is reasonable to assume that the equatorial solar wind co-rotates with the Sun up to the heliocentric distance where the solar wind kinetic energy density achieves its magnetic energy density (Parker 1958, 644-675) (Parker 1960, 821-866).

High-speed streams

Let us consider the formation of high-speed streams in the solar wind acceleration region owing to the interaction between the plasma and the magnetic field.

Coronograph images of the Sun reveal that structures of enhanced brightness of K-corona are organized in space as the ring engirdling the Sun. The band of coronal rays forms the upper rim of the ring. At the solar minimum, the ring is situated approximately around the equator (Hansen et al. 1974, 13-15) (Howard and Koomen 1974, 469-475) (Svalgaard et al. 1974, 157-172). The same ring is also seen on solar X-ray images: bright sources and regions without X-ray emission (so-called coronal holes) are the two extremes. Most coronal X-ray structures are seen as closed loops rooted in the chromosphere. It testifies for the confidence of coronal plasma inside magnetic field tubes. If the diffusion and the thermal conductivity of the coronal plasma are inhibited across the strong magnetic field, then the density and the temperature are increased in separate coronal regions. It leads to the intensification of electromagnetic emissions from these regions. Let us consider a part of the ball layer with the inner boundary at the coronal base ($r = R$) and the outer boundary at the sphere with the radius $r_{ex} \cong 5R$; the part is situated between the coaxial (with the solar rotation axis) cones $\theta = \theta_b$, $\theta = -\theta_b$; the regions of intensive electromagnetic emission of the given shapes $r = r(\theta,\varnothing)$ are eliminated from the part of the ball layer. The stationary transonic subalfvenic solar wind floods this separated space portion with open magnetic field lines. The high coronal temperature (~ 10^6 °K) smooths temperature gradients in the solar wind and allows assuming the isothermal flow in a zero-order approximation. The co-rotation of the subalfvenic solar wind in middle and low helio-latitudes at heliocentric distances closer than 5 R means that one may neglect the solar rotation influence there, i.e. one may put $Ro \rightarrow \infty$. The solar wind flow in this space portion is supposed to be directed along the magnetic field. The corresponding dimensionless stationary magneto-hydrodynamic equations are as follows:

$$\left(\vec{V} \cdot \nabla\right)\vec{V} + \frac{1}{\rho}\nabla P + \frac{1}{Fr}\frac{1}{r^2}\vec{e}_r + \frac{1}{M_A^2 \rho}\left[\vec{j} \times \vec{B}\right] = 0 \qquad (4.76)$$

$$\left(\nabla \cdot \rho\vec{V}\right) = 0 \qquad (4.77)$$

$$\left(\nabla \cdot \vec{j}\right) = 0 \qquad (4.78)$$

$$\left(\nabla \cdot \vec{B}\right) = 0 \qquad (4.79)$$

$$\left[\vec{V} \times \vec{B}\right] = 0 \qquad (4.80)$$

$$\left[\nabla \times \vec{B}\right] = \vec{j} \tag{4.81}$$

The Parker critical velocity $V^* = (2\kappa T^*/m_p)^{1/2}$ is chosen as the velocity unit, $Fr = 2\kappa T^* R/(GM_s m_p)$, M_A^2 is the square of the Alfven-Mach number (a parameter). The boundary conditions are as follows:

$$\rho(R,\theta,\varnothing) = \rho_s = const \tag{4.82}$$

$$\left.\frac{4\pi\rho V^2}{B^2}\right|_{r=R} = M_{As}^2 = const \tag{4.83}$$

$$B_\theta(r_{ex},\theta,\varnothing) = B_\varnothing(r_{ex},\theta,\varnothing) = 0 \tag{4.84}$$

$$B_\theta(r,\theta_b,\varnothing) = B_\theta(r,\pi-\theta_b,\varnothing) = 0 \tag{4.85}$$

$$B_r(R,\theta,\varnothing) = B_{rph}(R,\theta,\varnothing) \tag{4.86}$$

$$B_n(r,\theta,\varnothing)|_{r=r(\theta,\varnothing)} = 0 \tag{4.87}$$

Here M_{As} is the Alfven-Mach number at the coronal base. Boundary conditions (4.82), (4.83), (4.86) are related to the part of the coronal base outside of regions of the intense electromagnetic emission. Boundary conditions (4.82) and (4.83) are linked: if (4.82) is set, the magneto-hydrodynamic equations require setting (4.83). The boundary condition (4.87) means the absence of the magnetic field component normal to boundaries of the intense electromagnetic emission regions, which implicitly are assumed magnetically confined regions (remain immovable) so that the solar wind streamlines such obstacles. Boundary conditions (4.84)-(4.87) prohibit the solutions with open magnetic field lines in the absence of the current sheet above the neutral line of the photosphere magnetic field. It is impossible to specify the shape of the current sheet under the problem statement in case of the flow space symmetry absence. Therefore, let us follow Schatten (Schatten 1971, 232-246) and replace the boundary condition (4.86) as follows:

$$B_r(R,\theta,\varnothing)=\left|B_{rph}(R,\theta,\varnothing)\right| \tag{4.88}$$

Boundary conditions (4.84), (4.85), (4.87), (4.88) permit the existence of continuous solutions with open field lines while the current sheet is taken away owing to the replacement of the condition (4.86) by the condition (4.88) (Писанко 1985, 17-22). Magnetic field lines of such a solution are topologically equivalent to magnetic field lines of a "magnetic charge" situated "inside" the Sun (Fig. 4-1).

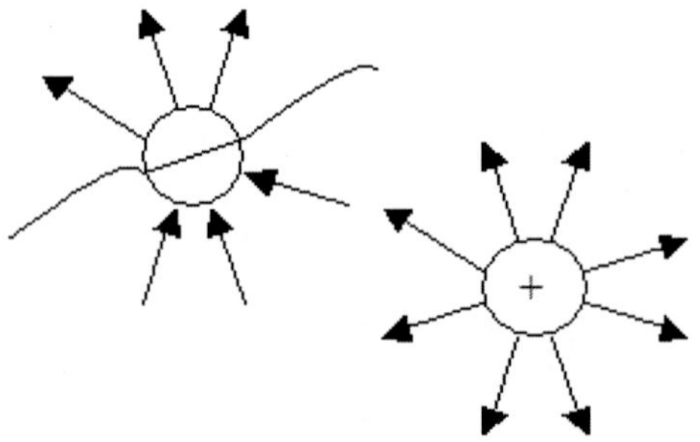

Fig. 4-1 Magnetic field line topologies of the solar wind (left) and a "magnetic charge" (right)

The substitution of Eq. (4.80) into Eq. (4.77), taking into account Eq. (4.79), gives:

$$\left(\vec{B} \cdot \nabla \frac{V}{B}\right) = 0 \tag{4.89}$$

The potential character of the gravitation, along with the barotropy of the flow, allows rewriting Eq. (4.76):

$$\nabla\left(\frac{V^2}{2} + \ln \rho - \frac{1}{Fr}\frac{1}{r}\right) = \left[\vec{V} \times \left[\nabla \times \vec{V}\right]\right] + \frac{1}{M_A^2 \rho}\left[\vec{j} \times \vec{B}\right] \tag{4.90}$$

The scalar multiplication of Eq. (4.90) by ***B*** (taking into account Eq. (4.80)) gives:

$$\left(\vec{B} \cdot \nabla\left(\frac{V^2}{2} + \ln \rho - \frac{1}{Fr}\frac{1}{r}\right)\right) = 0 \tag{4.91}$$

Integrating Eq. (4.89) and Eq. (4.91) along an open magnetic field line, taking into account boundary conditions (4.82), (4.83) and the condition $V|_{r=R} \ll 1$, one obtains the two integrals:

$$\frac{V^2}{2} + \ln \frac{\rho}{\rho_s} - \frac{1}{Fr}\left(1 - \frac{1}{r}\right) = 0 \tag{4.92}$$

$$\frac{\rho V}{B} = M_{As}\sqrt{\frac{\rho_s}{4\pi}} \tag{4.93}$$

which are valid in any point of the flow. The substitution of Eq. (4.93) into Eq. (4.92) gives the transcendental equation for V:

$$\frac{V^2}{2} - \ln V + F(M_{As}) = 0 \tag{4.94}$$

$$F(M_{As}) = \ln B + \ln\frac{M_{As}}{\sqrt{4\pi\rho_s}} + \frac{1}{Fr}\left(1 - \frac{1}{r}\right)$$

To determine the velocity V, one should find zeros of the function $f(V) = V^2/2 - \ln V + \mathrm{F}(M_{As})$ in the range $0 < V < +\infty$, where the function has its minimum $f_{\min} = f(1) = 0{,}5 + F(M_{As})$ at $V = 1$. Zeros exist in case of $f_{\min} \le 0$, i.e. when:

$$\ln B + \ln\frac{M_{As}}{\sqrt{4\pi\rho_s}} + \frac{1}{Fr}\left(1 - \frac{1}{r}\right) \le -\frac{1}{2} \tag{4.95}$$

Because

$$\ln B + \ln\frac{M_{As}}{\sqrt{4\pi\rho_s}} - \frac{1}{Fr}\frac{1}{r} \le \ln\frac{M_{As}}{\sqrt{4\pi\rho_s}} + \max\left\{\ln B - \frac{1}{Fr}\frac{1}{r}\right\},$$

the inequality (4.95) is valid for any value of M_{As}, satisfying the following inequality:

$$\ln\frac{M_{As}}{\sqrt{4\pi\rho_s}} + \max\left\{\ln B - \frac{1}{Fr}\frac{1}{r}\right\} \le -\frac{1}{2} - \frac{1}{Fr} \tag{4.96}$$

Hence, the sufficient condition of the solution existence is the limitation of the tolerance range of M_{As} to the range:

$$0 < M_{As} \le M_{Asc},$$

$$M_{Asc} = \sqrt{4\pi\rho_s}\exp\left(-\frac{1}{2}-\frac{1}{Fr}-\max\left\{\ln B-\frac{1}{Fr}\frac{1}{r}\right\}\right) \quad (4.97)$$

If $M_{As} > M_{Asc}$, then there are points in the space portion (for example, where the function $\ln B$ - $(r\,Fr)^{-1}$ reaches its absolute maximum), where the inequity (4.96) does not take place, and the function $f(V)$ has no zeros. In such points, the solution does not exist. If $M_{As} < M_{Asc}$, then at any point in the space portion $\mathrm{F}(M_{As}) < 1/2$. In this case, the function $f(V)$ always has two zeros – the first in the range $0 < V < 1$ while the second in the range $1 < V < \infty$; no points where $V = 1$ exist. Hence, possible is either the subsonic ($V < 1$) flow everywhere or the supersonic ($V > 1$) flow everywhere; there is no continuous junction between the subsonic regime and the supersonic one. If $M_{As} = M_{Asc}$, then points where $f_{\min} = 0$ ($V = 1$) exist in the space portion. In such points, one may write:

$$F(M_{Asc}) = \ln B + \ln\frac{M_{Asc}}{\sqrt{4\pi\rho_s}} + \frac{1}{Fr}\left(1-\frac{1}{r}\right) = -\frac{1}{2} \quad (4.98)$$

The substitution of Eq. (4.97) into Eq. (4.98) gives:

$$\frac{1}{r} = Fr\left(\ln B - \max\left\{\ln B - \frac{1}{Fr}\frac{1}{r}\right\}\right) \quad (4.99)$$

The heliocentric distance (dr) differentiation of Eq. (4.94) gives:

$$\left(V-\frac{1}{V}\right)\frac{dV}{dr}=-\frac{1}{B}\frac{dB}{dr}-\frac{1}{Fr}\frac{1}{r^2} \tag{4.100}$$

For $V = 1$, Eq. (4.100) requires:

$$-\frac{1}{B}\frac{dB}{dr}=\frac{1}{Fr}\frac{1}{r^2} \tag{4.101}$$

If B is a monotone decreasing function of the heliocentric distance ($dB/dr < 0$), Eq. (4.101) is as follows:

$$\frac{1}{B}\left|\frac{dB}{dr}\right|=\frac{1}{Fr}\frac{1}{r^2} \tag{4.102}$$

Substituting Eq. (4.99) into the right-hand side of Eq. (4.102), one obtains:

$$\frac{1}{B}\left|\frac{dB}{dr}\right|=Fr\left(\ln B-\max\left\{\ln B-\frac{1}{Fr}\frac{1}{r}\right\}\right)^2 \tag{4.103}$$

Because the left-hand side of Eq. (4.103) is a decreasing function of the heliocentric distance while the right-hand side of Eq. (4.103) is an increasing function of the heliocentric distance and both are continuous functions of the heliolatitude and the heliolongitude, the sufficient condition of the existence of the continuous junction between subsonic and supersonic regimes (a critical surface) is as follows:

$$\min\left\{\frac{1}{B}\left|\frac{dB}{dr}\right|\right\}>Fr\cdot\max\left\{\left(\ln B-\max\left\{\ln B-\frac{1}{Fr}\frac{1}{r}\right\}\right)^2\right\} \tag{4.104}$$

It is clear from the inequality (4.104) that if $Fr \rightarrow 0$, then the critical surface extends at infinity, while if $Fr > \max\{B|dB/dr|\text{-}1\}$, then the critical surface goes under the photosphere. Thus the isothermal transonic flow may exist for the Froude numbers belonging to the limited range. Similar to the Parker spherically symmetric isothermal solar wind (Parker 1958, 644-675), where the only parameter controls the continuous transonic flow – the coronal temperature value, the three-dimensional isothermal continuous transonic flow in the given magnetic field is controlled by the only parameter as well – the value of the function of function:

$$M_{Asc} = \sqrt{4\pi\rho_s}\exp\left(-\frac{1}{2}-\frac{1}{Fr}-\max\left\{\ln B-\frac{1}{Fr}\frac{1}{r}\right\}\right)$$

The substitution of Eq. (4.92) into Eq. (4.90) gives:

$$\left[\vec{V}\times\left[\nabla\times\vec{V}\right]\right]+\frac{1}{M_A^2\rho}\left[\vec{j}\times\vec{B}\right]=0 \qquad (4.105)$$

From Eq. (4.93) and Eq. (4.80), one obtains:

$$\vec{V} = M_{Asc}\sqrt{\frac{\rho_s}{4\pi}\frac{1}{\rho}}\vec{B} \qquad (4.106)$$

Substituting Eq. (4.106) into Eq. (4.105), one obtains after transformations:

$$\frac{\rho-\rho_s\dfrac{M_A^2M_{Asc}^2}{4\pi}}{M_{Asc}\rho M_A^2\sqrt{\dfrac{\rho_s}{4\pi}}}\left[\vec{j}\times\vec{B}\right]=\left(\vec{B}\cdot\nabla\frac{M_{Asc}\sqrt{\dfrac{\rho_s}{4\pi}}}{\rho}\right)\vec{B}- \tag{4.107}$$

$$B^2\nabla\frac{M_{Asc}\sqrt{\dfrac{\rho_s}{4\pi}}}{\rho}$$

The vector multiplication of Eq. (4.107) by ***B*** gives after transformations:

$$\vec{j}=\frac{\rho_A}{\rho(\rho-\rho_A)}\left[\vec{B}\times\nabla\rho\right] \tag{4.108}$$

Here $\rho_A = \beta M_{Asc}^2\rho_s/(4\pi)$ is the plasma density at the Alfven point (where kinetic and magnetic energies become equal); moreover, $\rho > \rho_A$ everywhere in the space portion; the latter is ensured by the appropriate choice of the Alfven-Mach number. Eq. (4.108) determines the density of transverse (to the magnetic field) electric currents needed to ensure the force balance in perpendicular (to the magnetic field) directions. The absence of field-aligned electric currents follows Eq. (4.78), Eq. (4.84), and Eq. (4.85). The condition of the absence of field-aligned electric currents is as follows:

$$\left(\vec{B}\cdot\left[\nabla\times\vec{B}\right]\right)=0 \tag{4.109}$$

Eq. (4.109) is satisfied identically when:

$$\vec{B} = \mu \nabla \Psi \tag{4.110}$$

Here $\mu = \mu(r,\theta,\varnothing)$, $\Psi = \Psi(r,\theta,\varnothing)$. At that:

$$\vec{j} = \left[\nabla \times \vec{B}\right] = \frac{1}{\mu}\left[\nabla \mu \times \vec{B}\right] \tag{4.111}$$

From Eq. (4.108) and Eq. (4.111), one obtains:

$$-\frac{\rho_A}{\rho(\rho - \rho_A)}\nabla \rho = \frac{1}{\mu}\nabla \mu \tag{4.112}$$

The solution of Eq. (4.112) is as follows:

$$\mu = C_6 \frac{\rho}{\rho - \rho_A} \tag{4.113}$$

One may put to unity the parameter C_6 = *const* without the loss of generality. Substituting Eq. (4.113) into Eq. (4.110), one obtains:

$$\vec{B} = \frac{\rho}{\rho - \rho_A}\nabla \Psi \tag{4.114}$$

The substitution of Eq. (4.114) into Eq. (4.79) gives:

$$\left(\nabla \cdot \frac{\rho}{\rho - \rho_A}\nabla \Psi\right) = 0 \tag{4.115}$$

It is a stationary diffusion equation with the variable coefficient $\rho(\rho\text{-}\rho_A)^{-1}$. Eq. (4.115) degenerates into the Laplace equation ($\rho >> \rho_A$, $\rho(\rho\text{-}\rho_A)^{-1} \rightarrow 1$) near the Sun. At the same time, Eq. (4.114) degenerates into $\boldsymbol{B} = \nabla\Psi$, i.e.

the magnetic field near the Sun proves to be almost potential. The deviation from potentiality increases with the heliocentric distance increase. The magnetic field for the given plasma density spatial distribution is determined by the Eq. (4.114). Here Ψ is the solution of the stationary diffusion equation (4.115) under the boundary conditions:

$$\Psi\left(r_{ex},\theta,\varnothing\right)=const \tag{4.116}$$

$$\left.\frac{\partial\Psi}{\partial r}\right|_{r=R}=\left|B_{rph}\left(R,\theta,\varnothing\right)\right| \tag{4.117}$$

$$\frac{\partial\Psi}{\partial n}=0 \tag{4.118}$$

$$\left.\frac{\partial\Psi}{\partial\theta}\right|_{\theta=\theta b}=\left.\frac{\partial\Psi}{\partial\theta}\right|_{\theta=\pi-\theta b}=0 \tag{4.119}$$

based on Eq. (4.114) and boundary conditions (4.84),(4.85),(4.87),(4.88) (Писанко, 1989). The condition (4.117) relates to the coronal base outside magnetically confined regions, but the condition (4.118) – to boundaries of magnetically confined regions.

The magneto-hydrodynamic equilibrium of the flow is calculated with the help of a two-step iteration procedure. The potential magnetic field (calculated from Eqs. (4.114)-(4.119)) is chosen as a zero-order approximation. The plasma density spatial distribution in the given magnetic field is calculated from Eqs. (4.92)-(4.93). Then a first-order approximation of the magnetic field in the given plasma density spatial distribution is calculated from Eqs. (4.114)-(4.119) and so on. The calculation continues until the changes from previous to subsequent iterations become small. After finishing the iteration procedure, the equilibrium solar wind open magnetic field line geometry is reconstructed. Directions of open magnetic field lines connected to the parts of the coronal base where $B_{rph}(R,\theta,\varnothing) < 0$ are changed to opposite directions so that $\boldsymbol{B} \rightarrow -\boldsymbol{B}$ in the space portion full with these lines. After such transformation, the "fictitious solar magnetic charge" (introduced by the boundary condition (4.88)) disappears, the total magnetic flux across the coronal base becomes zero, but at the same time, current sheets carrying surface currents $\boldsymbol{i} = 2[\boldsymbol{n}\times\boldsymbol{B}]$ appear above neutral lines of the photospheric magnetic field. The Rankine-Hugoniot conditions degenerate to tangential discontinuity conditions on current sheets. The balance of the total (thermal plus magnetic) pressure is ensured by the continuity of the plasma density and the magnetic field across the sheets due to the current sheet construction procedure. The electric current component normal to the sheet determines the surface electric current divergence on the current sheet.

This model allows calculating various scenarios of the formation of coronal roots of high-speed streams in the near-equatorial solar wind. The natural first step is to calculate situations with the simplest (but essentially three-dimensional) geometry of magnetically confined regions. As an example of the first in the world practice (1985) calculation of the three-dimensional solar wind taking into account the interaction between the plasma and the magnetic field in the flow formation region self-consistently, let us describe the results of such calculations in the presence of the band (gathered from helmet streamers) engirdling the Sun around the equator with northward and southward meanders (Писанко 1985, 371-375). The high-speed streams are formed in this case above solar regions with open magnetic configurations owing to the electric currents induced in the flow (Писанко 1986, 1-5). When the magnetic field decreases with the heliocentric distance, the role of electric currents induced in the solar wind increases and is manifested as the increase of magnetic spatial variations, that, in turn, increases spatial variations of magnetic tube cross-section areas, that, in turn, increases solar wind velocity spatial variations. The latter is known as the coronal roots of high-speed streams. Another interesting example is the calculation of the three-dimensional solar wind streamlining the ring, which engirdles the Sun around the equator with one helio-longitude cut of the given width (Pisanko 1994, 791-794). The cut models an equatorward extension of a polar coronal hole. The magnetic field at the coronal base is assumed constant for simplicity. The calculation shows that the velocity increase is formed in the supersonic flow in the region of the minimum magnetic field – the speed of supersonic flow is inversely proportional to the magnetic tube cross-section area. In the considered case, the minimum magnetic field is situated on the backside (to the cut) part of the solar equator ring. The formation of the reduced magnetic field in the solar wind above the region with an intense magnetic field confining the coronal plasma is possible when current sheets on its boundary screen the magnetically confined region; the higher the electric conductivity is, the more effective the screening that takes place. Curiously, this simple coronal configuration calculated in 1992 was realized in August-September 1996 in the solar activity minimum (Gibson et al. 1999, 871-879), and the high-speed stream was observed in the ecliptic on the backside to the equatorward extension of the north polar coronal hole in the absence of any equatorial coronal holes there. These examples testify to the magnetic field controller functions in forming coronal roots of high-speed streams: a magnetic field spatial distribution can be adjusted to form the high-speed stream above both a closed magnetic coronal structure and an open one.

The knowledge of the region of (5–20) *R* is incomplete. We know from radio astronomical observations that the magnetic field is almost radial in this region, at least at low and middle heliolatitudes (Уиндхэм 1968, 137-

150). The same conclusion can be drawn from solar eclipse observations of the coronal rays representing the magnetic field geometry (Несмянович 1971, 74-76) (Vsekhsvjatsky 1963, 129-136). Estimates of the Froude and the Rossby numbers show that the solar gravitation influence and the solar rotation one upon the flow dynamics can be neglected in this region in zero-order approximation ($Fr \to \infty$, $Ro \to \infty$). Let us describe the velocity and the magnetic field in the spherical coordinate system in the form $\boldsymbol{V}(r,\theta,\varnothing) = \{V_r(r,\theta,\varnothing), 0, 0\}$, $\boldsymbol{B}(r,\theta,\varnothing) = \{B_r(r,\theta,\varnothing), 0, 0\}$. In this case, stationary magneto-hydrodynamic equations degenerate as follows:

$$\frac{1}{r^2}\frac{\partial}{\partial r}\left(r^2 \rho V_r\right) = 0,$$

$$\frac{1}{r^2}\frac{\partial}{\partial r}\left(r^2 B_r\right) = 0,$$

$$\rho V_r \frac{\partial V_r}{\partial r} + \frac{1}{M^2}\frac{\partial P}{\partial r} = 0,$$

$$\frac{1}{r}\frac{\partial P}{\partial \theta} + \frac{1}{r}\frac{M^2}{M_A^2} B_r \frac{\partial B_r}{\partial \theta} = 0,$$

$$\frac{1}{r\sin\theta}\frac{\partial P}{\partial \varnothing} + \frac{1}{r\sin\theta}\frac{M^2}{M_A^2} B_r \frac{\partial B_r}{\partial \varnothing} = 0$$

After the integration, one obtains:

$$r^2 \rho V_r = r_0^2 \rho_0 V_{r0} \tag{4.120}$$

$$r^2 B_r = r_0^2 B_{r0} \tag{4.121}$$

$$P + \frac{M^2}{2M_A^2} B_r^2 = F(r) \tag{4.122}$$

$$\rho_0 V_{r0} \frac{r_0^2}{r^2} \frac{\partial V_r}{\partial r} + \frac{1}{M^2} \frac{\partial P}{\partial r} = 0 \tag{4.123}$$

Here $\rho_0(r_0,\theta,\varnothing)$ is the density, $V_{r0}(r_0,\theta,\varnothing)$ is the velocity, $B_{r0}(r_0,\theta,\varnothing)$ is the magnetic field (with zero average value over both heliolatitudes and heliolongitudes) at the boundary level $r_0 = 5R$; $F(r)$ is an arbitrary function of the heliocentric distance. Let us suppose that the furthest (from the photosphere) point of the sonic surface is situated closer to the Sun than 5 R, and the closest (to the photosphere) point of Alfven's surface is situated further from the Sun than 5 R (Parker 1958, 644-675) (Weber and Davis 1967, 217-227). To specify $F(r)$, the energy equation could be used, but it is difficult to write the credible one: the specific mode of the energy transmission in the solar corona remains largely unknown because of plasma waves, magnetic fields etc. We, therefore, use known properties of the flow to specify $F(r)$. According to Eq. (4.122), $F(r)$ is uniquely determined by the pressure profile $P_m(r,\theta_m,\varnothing_m)$ along the ray $\theta = \theta_m$, $\varnothing = \varnothing_m$ distinguished somehow among the others. From Eq. (4.121) and Eq. (4.122), the pressure is as follows:

$$P(r,\theta,\varnothing) = P_m(r,\theta_m,\varnothing_m) + \frac{M^2}{2M_A^2}\left(\frac{r_0}{r}\right)^4 \left[B_{rm}^2(r_0,\theta_m,\varnothing_m) - B_{r0}^2(r_0,\theta,\varnothing)\right] \tag{4.124}$$

Substituting Eq. (4.124) into Eq. (4.123) and integrating from r_0 up to a current point r, one can eventually obtain the expression for the flow velocity as follows:

$$V_r(r,\theta,\varnothing) = V_{r0}(r_0,\theta,\varnothing)H(r,\theta,\varnothing) \tag{4.125}$$

$$H(r,\theta,\varnothing) = 1 - \frac{1}{\rho_0 V_{r0}^2}\left(\frac{r}{r_0}\right)^2 \frac{1}{M^2}\int_{r_0}^{r}\frac{\partial P_m}{\partial \xi}d\xi + \left[1 - \left(\frac{r_0}{r}\right)^2\right]\frac{B_{rm}^2(r_0,\theta_m,\varnothing_m) - B_{r0}^2(r_0,\theta,\varnothing)}{\rho_0 V_{r0}^2 M_A^2}$$

From Eq. (4.120) and Eq. (4.125), one obtains the expression for the density as follows:

$$\rho(r,\theta,\varnothing) = \rho_0(r,\theta,\varnothing)\left(\frac{r_0}{r}\right)^2 \frac{1}{H(r,\theta,\varnothing)}$$

The pressure profile $P_m(r,\theta_m,\varnothing_m)$ can be specified in many ways. For example, direct measurements of the solar wind on "Helios 1-2" spacecraft indicate that the slow ($V < 400$ km s^{-1}) plasma cools nearly adiabatically (Schwenn et al. 1981, 126-130). Let us, therefore, assume that, at the given heliocentric distance, the slowest local element of the flow must cool adiabatically. Mathematically this means that for the ray with the coordinates $\theta = \theta_m$, $\varnothing = \varnothing_m$ where $V_{r0}(r_0,\theta_m,\varnothing_m) = \min\{V_{r0}(r_0,\theta,\varnothing)\} = V_{0m}$, the pressure and the density are adiabatically linked as follows:

$$\frac{V_{rm}^2}{2} - \frac{1}{M^2}\frac{5P_{0m}}{2\rho_{0m}}\left[1 - \left(\frac{P_m}{P_{0m}}\right)^{2/5}\right] = \frac{V_{0m}^2}{2} \tag{4.126}$$

$$r^2 \rho_m V_{rm} = r_0^2 \rho_{0m} V_{0m}$$

$$\frac{P_m}{P_{0m}} = \left(\frac{\rho_m}{\rho_{0m}}\right)^{5/3} \tag{4.127}$$

Here $V_{rm} = V_r(r,\theta_m,\varnothing_m)$, $\rho_m = \rho(r,\theta_m,\varnothing_m)$, $P_m = P(r,\theta_m,\varnothing_m)$, $\rho_{0m} = \rho(r,\theta_m,\varnothing_m)$, $P_{0m} = P(r,\theta_m,\varnothing_m) = (2\kappa T/m_p)\rho_{0m} = V_s^2\rho_{0m}$, $T = T(r,\theta_m,\varnothing_m)$ is the solar wind temperature. Introducing a new variable $x = (V_{rm}/V_{0m})^{2/3}$ and another Mach number $M_1 = V_{0m}/V_s$, one obtains from Eqs. (4.126)-(4.127) an algebraic equation for x:

$$x^4 - \left(1 + \frac{5}{M^2 M_1^2}\right)x + \frac{5}{M^2 M_1^2}\left(\frac{r_0}{r}\right)^{4/3} = 0 \tag{4.128}$$

The real solution of Eq. (4.128) with $x > 1$ (the supersonic flow) is as follows:

$$x = \left(1 + \frac{5}{M^2 M_1^2}\right)^{1/3} \sqrt{D(\lambda)}\left[1 + \sqrt{\frac{1}{\sqrt{D^3(\lambda)}} - 1}\right]$$

$$D(\lambda)=\frac{1}{2}\left[\sqrt[3]{1+C(\lambda)}+\sqrt[3]{1-C(\lambda)}\right], C(\lambda)=\sqrt{1+\lambda^3},$$

$$\lambda=-\frac{20}{3}\sqrt[3]{4\left(\frac{r_0}{r}\right)^4\frac{M^2M_1^2}{\left(5+M^2M_1^2\right)^4}}$$

With known x, it is easy to find from Eq. (4.126):

$$P_m=P_{0m}\sqrt{\left[1-\frac{M^2M_1^2}{5}(x^3-1)\right]^5} \tag{4.129}$$

Another way to specify P_m in the supersonic transalfvenic solar wind in middle and low heliolatitudes can be based on the total (thermal plus magnetic) pressure balance across the cone boundary of the polar solar wind. In this case, the pressure profile $P_m(r)$ is determined by Eq. (4.59), where transverse variations of the total pressure in the polar solar wind are assumed to be small ($a_m, b_m << a_0$).

So, constructed has been the three-dimensional analytical solution of magneto-hydrodynamic equations describing the radial supersonic transalfvenic solar wind mathematically compatible with the given radial spatially inhomogeneous magnetic field (Pisanko 1996, (3)61-(3)64).

It follows from Eq. (4.125) that, first, the main radial acceleration of the solar wind takes place below 20 R, and, second, latitudinal and longitudinal velocity variations increase with the heliocentric distance. The simple estimates illustrate these statements. Let us suppose:

$\min\{V_{r0}(5R,\theta,\varnothing)\} = V_{r0}(5R,\theta_{min},\varnothing_{min}) = 235 \text{ km s}^{-1}$

$\max\{V_{r0}(5R,\theta,\varnothing)\} = V_{r0}(5R,\theta_{max},\varnothing_{max}) = 245 \text{ km s}^{-1}$

$\rho_0(5R,\theta_{min},\varnothing_{min}) = m_p\ 0{,}75\times10^5 \text{ cm}^{-3}$

$\rho_0(5R,\theta_{max},\varnothing_{max}) = m_p\ 0{,}65\times10^5 \text{ cm}^{-3}$

$B_{r0}(5R,\theta_{min},\varnothing_{min}) = 0{,}075$ Gauss

$B_{r0}(5R,\theta_{max},\phi_{max}) = 0{,}070$ Gauss

These are typical numerical modelling results of velocity, density, and magnetic field variations at the $r_0 = 5\ R$ level obtained for the subalfvenic transonic solar wind (Pisanko 1994, 791-794). For $|\theta_{max}-\theta_{min}| \sim \pi/2$, $|\varnothing_{max}-\varnothing_{min}| \sim \pi$, $M = 1.5$ taking into account Eq (4.160), one calculates at the heliocentric distance 20 R:

$V_r(20R,\theta_{min},\varnothing_{min}) = 330 \text{ km s}^{-1}$

$V_r(20R,\theta_{max},\varnothing_{max}) = 850 \text{ km s}^{-1}$

These estimates show that any small regular latitudinal and/or longitudinal magnetic variation in the (5–20) R heliocentric distance interval, where the flow has transalfvenic character, causes the strengthening of the regular latitudinal and/or longitudinal solar wind velocity variation from the 5% amplitude at the 5 R level in the subalfvenic flow to the 100% amplitude at the 20 R level in the superalfvenic flow. The latter is typical for "monster" high-speed streams observed at the Earth orbit in solar minima (Hundhausen 1977, 226-329). If there is no latitudinal and/or longitudinal variation of the magnetic field at the 5 R level, the solar wind velocity's latitudinal and/or longitudinal variation does not increase significantly with the heliocentric distance. Hence, the magnetic field latitudinal and longitudinal inhomogeneities control the solar wind acceleration effectiveness – the "velocity latitudinal and/or longitudinal variation amplifier" exists between 5 R and 20 R due to the transalfvenic character of the flow. As it is clear from Eq. (4.125), the solar wind is accelerated with the different effectiveness at all heliolatitudes and heliolongitudes when $\min\{V_{r0}\}$ and $\max\{B_{r0}\}$ are achieved at the same point $\theta = \theta_m$, $\varnothing = \varnothing_m$ at the r = 5 R level. Otherwise, local areas in the flow could exist where the solar wind would be decelerated. Such zones might have been observed (Coles et al. 1991, 13849-13859).

Everything written above leads to dividing the corona into three parts: near-equatorial one and two polar ones. In turn, the near-equatorial corona can be divided into the subalfvenic transonic one situated below 5 R and the transalfvenic supersonic one situated above 5 R. The solar wind is supersonic and superalfvenic at heliocentric distances above 20 R. The field-aligned electric currents (induced due to the solar rotation and realized by supra-thermal tails of velocity distributions of electrons and ions) create the additional pressure gradient and the additional acceleration in the polar solar wind. The energy of this polar solar wind "accelerator" maintains the three-dimensional spatial structure of the near-equatorial transalfvenic supersonic solar wind, owing to the total (thermal plus magnetic) pressure balance at cone boundaries between polar and near-equatorial coronae. It ensures the existence of the "solar wind velocity's latitude and/or longitude variation amplifier" in the near-equatorial corona. Coronal roots of high-speed streams in the near-equatorial corona are formed in weak magnetic field regions. This construction is applicable for the solar wind spatial structure calculation. The result is shown in Fig.4-2.

It is clear from Fig.4-2 that such composite calculations correctly diagnose the solar wind spatial structure in the solar wind acceleration region. One can see also coronal images (white light, SOHO, LASCO) and UV (SOHO, EIT, 195 nm) for the period from August 10, 1996, till September 6, 1996, studied in detail in the framework of "Whole Sun Month" – the program of solar observations in the very solar minimum (Gibson et al. 1999, 871-879). It is seen that the high-speed stream was registered above the equatorial trunk of the north polar coronal hole, while another high-speed stream – on the backside of the Sun where no near-equatorial coronal holes were observed.

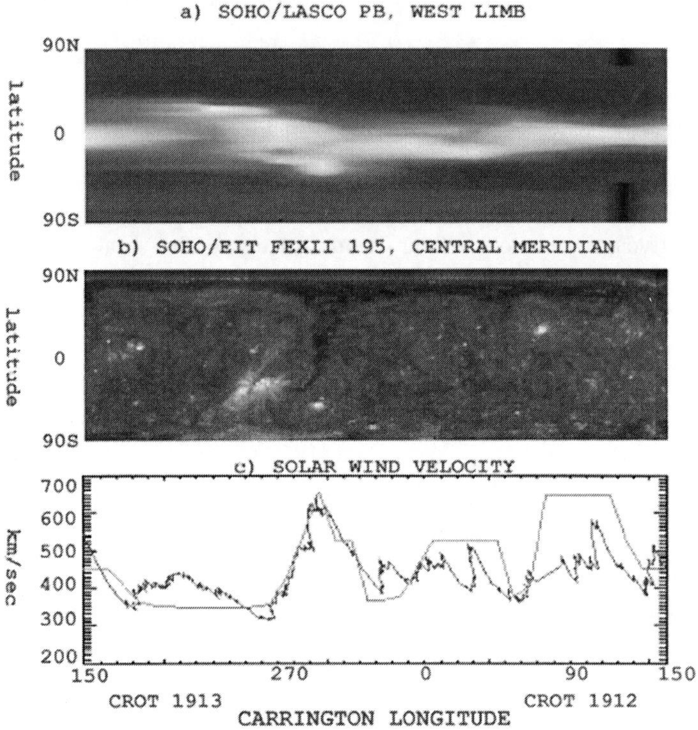

Fig. 4-2 The solar wind velocity calculated at $r = 20\ R$ (green line) at the ecliptic and direct measurements of the solar wind velocity at $r = 215\ R$ on the "Wind" spacecraft carried to $r = 20\ R$ (red line) (Gibson et al. 1999, 874)

An alternative approach to calculating the solar wind spatial structure provides the numerical integration of magneto-hydrodynamic equations using various finite difference techniques in the framework of limitations upon time and space steps imposed by the difference scheme theory. If boundary conditions do not depend on time, then stationary solutions can be obtained due to the evolution of arbitrary chosen initial conditions under $t \rightarrow \infty$. This so-called pseudo-viscosity method was used by Steinolfson et al. (1982, 730-742) to calculate the polytrope axially and equatorially symmetric magnetized non-rotating corona between the solar surface and the solar centred sphere of the radius 5 *R*. They chose the spherically symmetric Parker's flow (Parker 1958, 644-675) (Parker 1960, 821-866) and the dipole magnetic field at the initial moment. Their results were as follows. The magnetic field lines of the solution were closed near the equator and opened near poles. The upper rim of closed field lines was at the equator at the heliocentric distance 2 *R*. The velocity along closed field lines was almost zero, while there was transonic flow along open field lines. At the heliocentric distance 5 *R*, the pole flow velocity was higher than the equatorial one when the thermal pressure was higher than the magnetic one at the equatorial coronal base. Otherwise, the velocity was maximal at the latitude of 60°. Attempts were made to calculate in such a way the solar wind large scale structure corresponding to the photospheric magnetic field synoptic map for the given Carrington rotation (Usmanov 1993, 377-396). The photospheric magnetic field radial component map was reconstructed from the line-of-sight component synoptic observations at the inner boundary, while the density and temperature were chosen constant. At the initial moment, the spherically symmetric Parker's flow and the potential magnetic field fulfilling boundary conditions were chosen. Calculated velocities at the Earth orbit were significantly lower than observed ones on spacecraft. The introduction of hypothetical Alfven waves propagating from the solar convective zone increased calculated velocities (Usmanov 1996, 141-144); however, calculated latitudinal variations (~ 100 km s^{-1}) were smaller than observed ones (~ 300 km s^{-1}).

Coronal mass ejections

Let us consider physical scenarios of coronal mass ejections and the concept of the solar wind as an ensemble of many small coronal mass ejections.

In 1970th, white light corona observations on the "OSO 7" spacecraft registered a transient phenomenon (Tousey 1973, 713-730) which later (Hundhausen et al. 1984, 2639-2646) was called a "coronal mass ejection" – an observed bright feature of K-corona showing up in the coronograph field of view. Such features lasted from minutes to hours. The feature often looks like a bright external hoop (containing the dense coronal plasma) of the black cavity with a bright lump (strongly emitting in Hα line and interpreting, therefore, as the solar prominence plasma). Immediately after the birth, coronal mass ejections become parts of the solar wind. A coronal mass ejection brings 10^{12}-10^{13} kg of the solar matter into the interplanetary space so that the total solar wind mass flux increases up to 5%. The coronal mass ejection spatial scale is much less than the solar radius near the Sun. It rapidly increases when the ejection moves away into the interplanetary space and achieves tens of solar radii at the Earth orbit. Direct plasma measurements on spacecraft indicate that when the speed of the ejection exceeds the solar wind velocity on the value higher than the local sonic speed, the shock wave is observed ahead of the ejection. The coronal mass mainly ejects from the rim of the ring, engirdling the Sun at the equator; therefore, many ejections are oblate-shaped to the ecliptic in solar minima (Mulligan et al. 1998, 2959-2962) and move radially from the Sun in the solar wind background flow which is faster in high heliolatitudes. Suppose the coronal mass ejection spatial scale is comparable with the spatial scale of the solar wind velocity latitudinal variation. In that case, the ejection can achieve high heliolatitudes owing to the Magnus force (Moussas et al. 1995, 46). Because the ring locally loses its stability at the moment of the coronal mass ejection (McAllister and Hundhausen 1996, 171-184), a significant part of previously closed magnetic field lines could become open field lines. In the extreme, when all closed field lines become open ones (with the formation of the current sheet extending from the solar surface to infinity), the magnetic energy of the near solar environment can be enhanced after the coronal mass ejection (Aly 1984, 349-362) (Aly 1991, L61-L64).

Let us consider a coronal arc. On images of coronal arcs obtained on the "Trace" spacecraft, it is seen that the arc cross-section area does not increase with the height above the solar surface. If the arc were the plasma inside the potential (current-free) magnetic field tube, the tube cross-section area would increase with the height above the solar surface. Hence, electric currents exist in the coronal arc. Because the magnetic energy substantially exceeds the plasma thermal energy, the arc equilibrium is determined by the Ampere force, which is the sum of the magnetic pressure and the magnetic stress. The magnetic field is most significant at the arc axis; therefore, the magnetic pressure gradient pushes the arc to the neighboring weaker field regions to increase the arc radius. At the same time, the magnetic tightening force (decreasing the length of the magnetic field line) impedes the arc expansion into neighboring weaker field regions. The density of the resultant Ampere hoop force per unit of current length acting to increase the arc radius is as follows (Chen 1989, 453-470):

$$F_h = \frac{I_t^2}{cA}\left[\ln\left(\frac{8A}{a}\right) - \frac{1}{2}\frac{B_t^2}{B_p^2} - 1 + \frac{B_t a^2}{2I_t}\right]$$

Here I_t is the electric current along the arc axis, A is the big arc radius, a is the small arc radius, B_t is the magnetic field component along the arc axis (along the circle of the big radius A of the arc), B_p is the magnetic field component along the circle of the small radius (a) of the arc. It is supposed that a coronal mass ejection occurs when the magnetic pressure gradient exceeds the magnetic tightening force, and the unbalance increases with time. The nature of the coronal mass ejection energy is magnetic, and this magnetic energy is stored in the corona in the form of electric currents. The free magnetic energy is accumulated in an arc when the current flowing in the arc increases with time. The wriggle instability developing in the arc during its slow lifting under the action of Ampere's hoop force could heat the plasma inside the arc. It leads to the loss of the total pressure balance on the lateral arc surface at some critical height so that the arc cross-section area abruptly widens. The arc cross-section expansion leads to the plasma density decrease inside the arc. The impulse of the buoyancy force (owing to the density decrease) abruptly pushes the arc away from the Sun, creating the coronal mass ejection (Kuznetsov and Hood 1997, 61-80) (Kuznetsov and Hood 2000, 539-542).

The gradual increase of electric currents inside an arc could be initiated by motions of the arc magnetic tube legs frozen into the photosphere following photospheric plasma motions. If, during such wanderings, magnetic tubes of opposite polarities meet each other (near neutral lines of the photospheric magnetic field), local reconnections could occur. It leads to the formation of the current local sheet spreading at the top of the arc. When the current density exceeds the tearing instability threshold, the current sheet tears and a coronal mass ejection will be born.

Another scenario is also elaborated. It assumes an impulsive energy input from an external source into the arc. For example, it could be an abrupt shift of the arc magnetic tube legs in the photosphere. Such a shift twists the magnetic field around the arc axis, leading to the abrupt increase of the electric current along the arc axis. If the shift is sufficiently quick, then a coronal mass ejection occurs (Klimchuk 1990, 745-754).

The emergence of a new arc under an already existing one leads to the total magnetic flux increase in the region and could initiate a coronal mass ejection. A chromospheric flare could also abruptly increase the thermal pressure inside the arc so that the plasma will escape in the form of a coronal mass ejection. Observations show that flares and coronal mass ejections often co-occur (Delannee et al. 2000, 725-742) (Dryer 1996, 421-429).

The smaller the coronal mass ejection spatial scale is, the more often such small ejections are observed. The resolution of out-of-eclipse coronographs is limited, so the coronograph cannot register too small coronal mass ejections, which are supposed to be too many. During random wanderings of photospheric legs of coronal magnetic field lines of opposite polarities (Parker 1983, 642-647), reconnection is possible; therefore, a small coronal mass ejection (a plasmoid) could be pushed away from the Sun with the initial velocity, which is close to the local Alfven's speed ~ 1000 km s^{-1}. Schluter (Schluter 1957, 356-357) and Pneuman (Pneuman 1983, 468-482) drew attention to the specific acceleration mechanism of the already existing plasmoid during its motion in the spatially inhomogeneous background magnetic field (so-called "melon seed mechanism"). According to the mechanism, the plasmoid moving from the Sun is under the accelerating Ampere's force directed to the weaker background magnetic field region (so-called "mirror force"). The force is equal to $V_A^2 M_{pl}/L$, where V_A is the local Alfven speed, M_{pl} is the plasmoid mass, L is the spatial character scale of the background magnetic field inhomogeneity. Besides the mirror force, the drag acts on the spherical plasmoid moving in the background plasma. The drag is equal to $C\rho(r)V^2\pi a^2$ where V is the plasmoid velocity, a is the plasmoid radius, $\rho(r)$ is the background plasma density, C = 0.1–1 is the drag constant. The solar gravitation also influences the plasmoid motion; however, the plasmoid speed of about 1000 km s^{-1} exceeds the solar escape velocity; therefore, the solar gravitation can be neglected in a zero-order approximation. When the drag balances the mirror force, the plasmoid velocity is as follows (Pneuman 1983, 468-482):

$$V = V_A \sqrt{\frac{4a}{3CL} \frac{d_{in}}{d_{out}}}$$

Here d_{in} is the plasma density inside the plasmoid, d_{out} is the background plasma density (outside the plasmoid). At the coronal base near the reconnection region, $a = L$ and the plasmoid velocity can be estimated as 600–2000 km s^{-1}. On a very short boost part of the trajectory, the plasmoid gets an initial velocity of the local Alfven's speed in the reconnection region where magnetic fields are highly inhomogeneous. Then the melon seed mechanism continues the plasmoid acceleration not so intensively but at a much longer part of its trajectory. During its travel away from the Sun, the plasmoid enlarges tens or even one hundred times already at heliocentric distances (2–4) *R* (Pneuman and Cargill 1985, 653-664). However, not all plasmoids achieve the heliocentric distance of even 2 *R*: enlarging plasmoids fulfill all available space below 2 *R* so that plasmoids containing magnetic fields of opposite polarities could reconnect and create both new plasmoids and Alfven's waves. Such continuously enlarging plasmoids form the inhomogeneous plasma medium where the melon seed mechanism additionally accelerates them. One may suppose that the solar wind in its acceleration region is such a medium of small plasmoids – the maximum acceleration occurs at the heliocentric distance of the maximal turbulence. This scenario implies that the solar wind derives its momentum and energy on small spatial scales of the order of a nano-flare in the form of small scale plasmoids. During its motion away from the Sun, plasmoids enlarge and thereupon, the energy transfers along the spectrum of turbulence via the "reversed cascade" from small scales to large ones (Mullan 1990, 520-535), not via the "direct cascade" from large scales to small ones. Plasmoids pick up the background coronal plasma, and the added mass grows with the heliocentric distance. It could be why the solar wind velocity lower than 200–250 km s^{-1} has never been registered. Similar pick-up processes were observed, particularly in experiments with artificial injections of gases into the solar wind.

Coronal mass ejections demonstrate accelerated motion up to heliocentric distances (2–4) R and move mechanically with almost constant speed at more considerable distances. At the Earth orbit, a coronal mass ejection is the colossal structure of about 0.3 AU in across-track direction containing the magnetic field of 18 nT on average that is several times higher than the average interplanetary magnetic field value of 5 nT at the Earth orbit, and the magnetic pressure inside an ejection is significantly higher than the thermal one. In such a situation, the magnetic field inside an ejection (moving mechanically) can be modeled by the force-free magnetic field of the cylinder of infinite length (Burlaga 1988, 7217-7224).

Plasma polarization electric field in the acceleration region

Let us calculate the plasma polarization electric field heliocentric distance dependence in the solar corona, which corresponds to the empirical approximation of the "solar wind acceleration curve", derived from solar corona radio-sounding experiments (Pisanko and Yakovlev 2020, 1048-1053). A few such experiments were conducted under the minimum to moderate solar activity to determine the solar wind velocity heliocentric distance dependence in the solar wind acceleration region. However, velocity values obtained at the same heliocentric distance but different periods may differ from each other sometimes significantly; the difference is conditioned on both velocity variations and measurement errors. All available experimental values determined the averaged heliocentric distance dependence of solar wind velocity. Individual experimental velocity values were averaged over heliocentric distance intervals of the two solar radii length; each such interval contained from 6 to 12 velocity measurements. Obtained this way, the averaged heliocentric distance dependence of the solar wind velocity (Yakovlev and Pisanko 2018, 552-566) is presented by rectangles in Fig. 4-3.

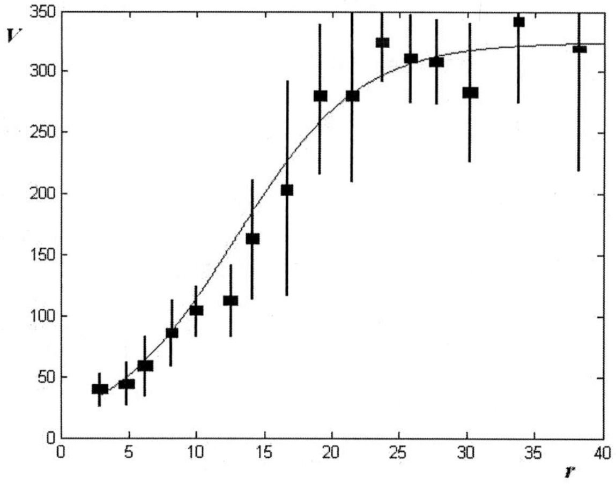

Fig. 4-3 The heliocentric distance dependence of the solar wind velocity; the heliocentric distance (abscissa axis) r is measured in solar radii, while the solar wind velocity (ordinate axis) V – in km s^{-1}; averaged (over heliocentric distance intervals of the two solar radii length each) experimental velocity values are shown by rectangles; the rms error bar of each average value is shown by vertical line segment; the approximation is shown by a solid line (Pisanko and Yakovlev 2020, 1049)

The averaged heliocentric distance dependence of the solar wind velocity is described by the approximation (Yakovlev and Yakovlev 2019, 237-242), which we rewrite in the form:

$$V(r) = \frac{V_0}{1 + e^{e - \delta r}} \qquad (4.130)$$

Here e=2.7183 is Euler's number, and the parameter values are as follows V_0=325 km s^{-1}, δ=0.2. Eq. (4.130) approximates the solar wind acceleration curve between 3 and 20 solar radii. The heliocentric distance r in Eq. (4.130) is measured in solar radii, while the solar wind velocity V –

in km s^{-1}. The solid line shows the approximation in Fig. 4-3. It corresponds to averaged experimental data within the limits of ± 20%.

Let us now demonstrate how to reconstruct the plasma polarization electric field heliocentric distance dependence in the solar wind acceleration region from known empirical heliocentric distance dependence of the solar wind velocity. As the solar gravitation confines coronal protons but cannot confine coronal electrons, the Coulomb attraction between protons and electrons leads to the creation of the potential electric field of the coronal plasma polarization, which prevents the free escape of electrons from the solar gravitational well, but obliges them to tow protons into the interplanetary space. In the case of the static corona, the plasma polarization electric field is known as the Pannekoek-Rosseland field (Pannekoek 1922, 107-118) (Rosseland 1924, 720-729). In the framework of the exospheric approach, the given empirical acceleration curve of the solar wind opens an opportunity to calculate a semi-empirical heliocentric distance dependence of the plasma polarization electric field, which corresponds to the empirical acceleration curve. At the same time, there is no need to define the solar wind electron component model concretely. It is enough to consider only solar wind protons. One may assume collisionless protons above the exobase, which are under the action of the two forces – the gravitational force from the

Sun and the electric force from solar wind electrons. The spherical symmetry of the problem is defined by the limitations of the empirical acceleration curve construction procedure. The motion equation of collisionless solar wind protons

$$m_p nV\frac{dV}{dr} = -G\frac{m_p nM_s}{r^2} + qE \qquad (4.131)$$

along with the Coulomb law (the Gauss theorem)

$$\frac{1}{r^2}\frac{d}{dr}\left(r^2 E\right) = \frac{q}{\varepsilon_0} \qquad (4.132)$$

under the constancy of the solar wind mass flow

$$nVr^2 = K = const \qquad (4.133)$$

lead (by the substitution of Eq. (4.132) and Eq. (4.133) to Eq. (4.131)) to an equation:

$$m_p K\frac{dV}{dr} = -G\frac{m_p M_s K}{Vr^2} + \varepsilon_0 E\frac{d}{dr}\left(r^2 E\right) \qquad (4.134)$$

Here m_p is the proton mass, n is the concentration of protons, V is the solar wind velocity, r is the heliocentric distance, G is the gravitational constant, M_s is the solar mass, q is the electric charge density, E is the polarization electric field intensity, ε_0 is the permittivity of vacuum, K is the solar wind mass flow. The value of the solar wind mass flow K is (for distinctness) taken from (Feldman et al. 1977, 353-354). Let us choose V_0=325 km s^{-1}

as the unit of velocity, the heliocentric distance to exobase r_{ex} as the unit of distance, the electric field intensity E_0 at the exobase as the unit of polarization electric field intensity, and introduce dimensionless variables as well as dimensionless parameters

$$N_1 = \frac{2m_p K V_0}{\varepsilon_0 E_0^2 r_{ex}^2}$$

$$N_2 = \frac{2m_p K G M_s}{\varepsilon_0 E_0^2 V_0 r_{ex}^3}$$

Eq. (4.134) in dimensionless variables is as follows:

$$N_1 r^2 \frac{dV}{dr} = -\frac{N_2}{V} + \frac{d}{dr}\left(r^2 E\right)^2 \qquad (4.135)$$

The dimensionless parameters are, correspondingly, the ratio of solar wind kinetic energy density to polarization electric field energy density at the exobase level (N_1) and the ratio of solar wind potential energy density to polarization electric field energy density at the exobase level (N_2). Let us specify the empirical heliocentric distance dependence of the solar wind velocity according to Eq. (4.130), which (in dimensionless variables) is as follows:

$$V = \frac{1}{1 + e^{e-\beta r}} \qquad (4.136)$$

Here $\beta = \delta r_{ex}/R$, R is the solar radius. The substitution of Eq. (4.136) to Eq. (4.135) gives:

$$\frac{d}{dr}\left(r^2 E\right)^2 = N_1 \beta \frac{r^2 e^{e-\beta r}}{\left(1+e^{e-\beta r}\right)^2} + N_2\left(1+e^{e-\beta r}\right) \tag{4.137}$$

Integrating Eq. (4.137) from the exobase to the current heliocentric distance gives:

$$\left(r^2 E\right)^2 = 1 + N_1 \beta \int_1^r \frac{r^2 e^{e-\beta r}}{\left(1+e^{e-\beta r}\right)^2} dr + N_2 \int_1^r \left(1+e^{e-\beta r}\right) dr \tag{4.138}$$

One obtains after the integration:

$$\int_1^r \left(1+e^{e-\beta r}\right) dr = r - 1 + \frac{1}{\beta}\left(e^{e-\beta} - e^{e-\beta r}\right) \tag{4.139}$$

$$\int_1^r \frac{r^2 e^{e-\beta r}}{\left(1+e^{e-\beta r}\right)^2} dr = \frac{e^2}{\beta^3}\left[arctg\left(e^{e-\beta}\right) - arctg\left(e^{e-\beta r}\right)\right]$$

$$+\frac{2e}{\beta^3}\left[\ln\frac{1+e^{e-\beta}}{1+e^{e-\beta r}}+(e-\beta r)\frac{e^{e-\beta r}}{1+e^{e-\beta r}}-(e-\beta)\frac{e^{e-\beta}}{1+e^{e-\beta}}\right]$$
$$+\frac{1}{\beta^3}\left[\frac{(e-\beta r)^2}{1+e^{e-\beta r}}-\frac{(e-\beta)^2}{1+e^{e-\beta}}-\beta^2\left(r^2-1\right)+2e\beta(r-1)\right]$$
$$+\frac{4}{\beta^3}\left[Li_2\left(1+e^{e-\beta}\right)-Li_2\left(1+e^{e-\beta r}\right)\right] \tag{4.140}$$
$$+\frac{2}{\beta^3}\left[Li_2\left(\frac{1}{1+e^{e-\beta}}\right)-Li_2\left(\frac{1}{1+e^{e-\beta r}}\right)\right]$$
$$+\frac{1}{\beta^3}\left[\ln^2\left(1+e^{e-\beta}\right)-\ln^2\left(1+e^{e-\beta r}\right)\right]$$

Here Li_2 is the dilogarithm function (Stegun 1972, 998-1010). Eqs. (4.138)-(4.140) allow calculating the polarization electric field heliocentric distance dependence. With known heliocentric distance dependence of the plasma polarization electric field, it is easy to calculate the heliocentric distance dependence of the electric charge density. From Eq. (4.135), one obtains:

$$\frac{1}{r^2}\frac{d}{dr}\left(r^2E\right)=\frac{1}{2r^2E}\left(N_1\frac{dV}{dr}+\frac{N_2}{r^2V}\right) \tag{4.141}$$

Eq. (4.141) along with Eq. (4.132) and Eq. (4.136) allows determining the heliocentric distance dependence of the electric charge density q in the solar wind acceleration region as follows:

$$q = \frac{1}{2r^2 E}\left(N_1 \frac{dV}{dr} + \frac{N_2}{r^2 V} \right) \qquad (4.142)$$

The dimensionless q should be multiplied by $\varepsilon_0 E_0 / r_{ex}$ to express the electric charge density in the SI unit system.

Let us estimate the polarization electric field intensity at the exobase. From Fig.4.3, one can see that the solar wind velocity in the base of the solar wind acceleration region is relatively low; for distinctness, let us suppose that the polarization electric field intensity at the exobase is of the order of the Pannekoek-Rosseland field intensity (i.e. is not differ dramatically from the polarization electric field intensity of the static corona). In the SI unit system, the latter is as follows (Pannekoek 1922, 107-118) (Rosseland 1924, 720-729):

$$E_0 = \frac{m_p G M_s}{2 q_e (r_{ex} R)^2} = 1.42856 \times 10^{-6} \frac{1}{r_{ex}^2} (V \cdot m^{-1}) \qquad (4.143)$$

The heliocentric distance to the exobase r_{ex} in Eq. (4.143) is dimensionless and is measured in solar radii, while q_e=1.6×10^{-19} C is an elementary electric charge. Therefore, the order of magnitude of the electric charge density in the solar wind acceleration region is $\varepsilon_0 E_0 / R \sim 10^{-25}$ C m^{-3}. It means that the electric charge density is negligible in the solar wind acceleration region.

The calculated heliocentric distance dependence of the polarization electric field for the exobase situated at three solar radii is shown in Fig. 4-4.

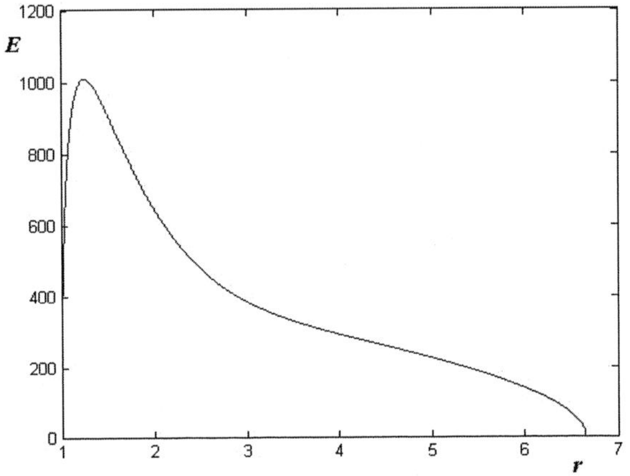

Fig. 4-4 The heliocentric distance dependence of the polarization electric field intensity if the electric field boundary value (E_0) at the exobase is equal to Pannekoek-Rosseland's and the heliocentric distance of the exobase is equal to three solar radii; the field intensity (the ordinate axis) E is expressed in V m^{-1}, while the unit on the abscissa axis r is equal to three solar radii (Pisanko and Yakovlev 2020, 1052)

One can see that the electric field intensity increases very rapidly with the heliocentric distance, achieves its maximum (~ 1000 V m^{-1}) at approximately 3.7 solar radii and then decreases to almost zero at approximately 20 solar radii. The maximum of the electric field intensity (situated not far from the inner boundary of the solar wind acceleration region) is connected with the need for solar wind protons to overcome the solar gravitation when they follow solar wind electrons on their way from the Sun keeping up the given empirical acceleration curve. The shift of the exobase closer to the Sun (to the heliocentric distance of two and a half solar radii) does not change the general character of the electric field intensity heliocentric distance dependence. The electric field maximum intensity becomes stronger (~ 1400 V m^{-1}) and is situated closer to the Sun (approximately at three solar radii). A more vigorous maximum electric field intensity is needed to overcome the stronger solar gravitation when keeping up the given empirical acceleration curve. The shift of the exobase further from the Sun (to the heliocentric distance of three and a half solar radii) also does not change the general character of the electric field intensity heliocentric distance dependence. However, the electric field maximum intensity becomes weaker (~ 800 V m^{-1}) and is situated further from the Sun (approximately at four and a third solar radii). The weaker maximum is sufficient to overcome the weaker solar gravitation when keeping up the given empirical acceleration curve. From Fig. 4-4, it is clear that the polarization electric field vanishes at heliocentric distances beyond approximately twenty solar radii where the solar wind moves mainly mechanically without the acceleration, the solar gravitation influence is negligible, so the substantial polarization electric field beyond the outer boundary of the solar wind acceleration region is not needed.

Let us estimate the relative deposit of the electric induction field (connected with the solar rotation) to the total electric field in the solar wind acceleration region. In the framework of the Parker spherically symmetric model of the interplanetary magnetic field, the induction field E_θ (directed along a heliolongitude circle) is (in the SI unit system) as follows:

$$E_\theta = \omega(r - R)B_s \frac{R^2}{r^2} \tag{4.144}$$

Here $\omega = 2.69\times10^{-6}$ s^{-1} is the angular velocity of solar equatorial regions, and $B_s \cong 10^{-4}$ T is the solar general magnetic field value. From Eq. (4.144), the induction field intensity is estimated as ~ 10^{-2} V m^{-1} in the solar wind acceleration region. It is substantially smaller than the characteristic values of the polarization field. It justifies neglecting the induction field when considering the solar wind acceleration region electric field in middle and low heliolatitudes.

The electric charge density heliocentric distance dependence given by Eq. (4.142) makes it easy to calculate the total positive electric charge of the solar wind acceleration region. In the spherically symmetric case presented in Fig. 4-4, the charge is equal to approximately 6500 Coulomb. It is interesting to compare this value with the absolute value of the total negative electric charge of the Earth. The latter is equal to approximately

500000 Coulomb and is one hundred times as much as the absolute value of the total positive electric charge of the solar wind acceleration region. At the same time, let us note that the value of 6500 Coulomb is one hundred times as much as the very crude estimate (~ 77 Coulomb) of the Sun's global positive electric charge (Neslusan 2001, 913-915).

Summary of Chapter 4

The electron thermal velocity under coronal temperatures is several times higher than the solar escape velocity, while the proton one is several times lower than the escape velocity. As the solar gravitation confines coronal protons but cannot confine coronal electrons, the Coulomb attraction between protons and electrons leads to the creation of the potential electric field of the coronal plasma polarization, which prevents the free escape of electrons from the solar gravitational well, but obliges them to tow protons into the interplanetary space. So, the coronal plasma evaporation explains the solar wind phenomenon by the difference between proton and electron masses under the equality of its electric charges.

Single fluid magneto-hydrodynamics is probably the simplest mathematical construction that describes self-consistently the interaction between the plasma and the magnetic field in the solar corona. The interaction is responsible for the inhomogeneous spatial structure of the acceleration region. The solar rotation induces electric currents in the solar wind and influences this way on its dynamics. The theory does not prohibit the existence of double electric layers in polar coronae. Above a double electric layer, field-aligned electric currents could exist in the polar solar wind. The carriers for these currents could be supra-thermal electrons and protons registered in the polar solar wind. These supra-thermal particles create an additional pressure gradient that additionally accelerates the polar solar wind. The requirement of the absence of the surface electric charge at the lateral boundary of the polar solar wind (which is treated as a current sheet) is ensured when, first, the equatorial corona rotates faster than the polar one and, second, the solar wind magnetic field on the "equatorial side" of the lateral boundary is inclined to the west (in anti-Parker's direction) near the Sun. Interestingly, the average incline of the photosphere magnetic field determined from synoptic maps for the period 1977-1992 proved to be 0.6° to the west. The solar wind, in turn, slows the solar rotation down on the cosmogonic time scale. Magneto-hydrodynamic calculations show that the magnetic field controls the formation of coronal roots of high-speed streams in the transonic subalfvenic solar wind: a high-speed stream can be formed above both the closed magnetic coronal structure and the open one, depending on the resulting self-consistent spatial structure of the open magnetic field in the solar wind. Further from the Sun in the supersonic transalfvenic solar wind, small magnetic variations lead to the amplification of spatial scale velocity variations from 5% at the heliocentric distance of 5 *R* up to 100% at the heliocentric distance of 20 *R*. In the transalfvenic solar wind, the magnetic field serves as a controller of the effectiveness of the solar wind acceleration at various heliolatitudes and heliolongitudes. The total (thermal plus magnetic) pressure balance across the lateral boundary of the polar solar wind maintains this inhomogeneous spatial structure of the near-equatorial solar wind, which is manifested as high-speed streams.

One may suppose that the solar wind in its acceleration region is a medium of very small coronal ejections – the maximum acceleration occurs at the heliocentric distance of the maximal turbulence. This scenario implies that the solar wind derives its momentum and energy on small spatial scales of the order of a nano-flare in the form of small scale coronal mass ejections. During its motion away from the Sun, these ejections enlarge and thereupon, the energy transfers along the spectrum of turbulence via the "reversed cascade" from small scales to large ones (not via the "direct cascade" from large scales to small ones). Coronal mass ejections pick up the background coronal plasma, and the added mass grows with heliocentric distance. It could be why the solar wind velocity lower than 200–250 km s^{-1} has never been registered.

The main result of chapter 4 is that both high-speed streams and coronal mass ejections are the manifestations of the free magnetic energy (stored in coronal electric currents) transfer into the kinetic energy of solar wind large scale structures.

CHAPTER 5

ASPECTS WE STILL DO NOT KNOW ABOUT THE SOLAR WIND

When analyzing the solar wind acceleration region, one should consider the physical conditions near the Sun where the solar wind originates and the conditions at the solar system's outskirts where the solar wind resolves in the local interstellar medium. These conditions and their reasons are poorly known. For example, we do not know why the solar corona is heated up to a million degrees, although the photosphere situated closer to the solar core is several hundred times cooler. We do not know why the solar wind exists in general: are there the solar-like stars where not the solar wind but accretion (the gravitation fall of the surrounding atmosphere to a star) is realized. We will discuss these and a few other unsolved questions in this last chapter of our book. We understand that we enter the hypothesis area, but we hope that the next generation will approach the solution. We finish the chapter describing the perspectives of solar wind forecasting and flights to the Sun.

Why does solar wind exist?

The problem of the solar wind origin emerged more than fifty years ago when the permanent nature of supersonic gas flows from the Sun became clear little by little due to ground-based observations of comets, cosmic ray variations, and geomagnetic disturbances. The first ambiguity of researchers was expressed in a simple question: why did the solar wind exist in general? The question has no sufficiently deep full answer yet, despite the considerable progress in exploring specific solar wind properties with the help of various spacecraft for more than fifty years. The considerable proportion of the remaining ambiguity is stipulated not only by the lack of objective information but also by the subjective view. The latter is connected, in particular, with the widespread exploitation of the standard model of the Sun. The model neglects the tiny structure details of the solar atmosphere – the manifestations of the absence of the mechanic and thermodynamic equilibrium. The solar wind is among such manifestations. Therefore, the answer to the question should be searched outside the framework of the standard model of the Sun. It seems that the problem cannot be solved purely theoretically. New observational information is needed and not only about the Sun. It is unclear why the plasma environment leaves the Sun but is not gathered near the Sun due to the gravitational accretion from the partly ionized local interstellar medium. It is unknown when and how the transformation from the gravitational accretion to the thermal expansion of the solar atmosphere at an early stage of the solar evolution occurred. The scientific literature (see the review (McKee and Ostriker 2007, 565-687)) does not answer. The details of such a transformation remain enigmatic though the nuclear fusion beginning inside the star could initiate the transformation. In 1950th, accretion was assumed as a heating mechanism for the solar corona. The physical idea looked attractive and straightforward: the tenuous interstellar gas fell to the Sun due to the gravitation, broke at the solar surface and spent its kinetic energy on the coronal heating. A polytrope accretion model proposed by Bondi (1952, 195-204) created, in fact, the mathematical foundation for the spherically symmetric polytrope solar wind model proposed by Parker because the Bernoulli equation (exploited in both models) contains the square of velocity. Therefore, every accretion solution (the negative sign of the radial velocity) corresponds to the entirely analogous expansion solution (the positive sign of the radial velocity) – the problem has degenerated. A similar but much more complicated situation takes place in exospheric models of the solar wind in the framework of collision-less kinetics. In this approach, the particle trajectories are divided into the outgoing, arriving, and finite ones. The latter, in principle, could not cross the solar surface. It creates the arbitrariness in specifying boundary conditions on characteristics not crossing the solar surface – virtual electrodes may occur in the solar corona; the degree of non-locality is similar to that in the Langmuir probe theory. Few analytical and numerical kinetic solutions were constructed and analyzed in the simplest case of the absence of finite particles. The results demonstrated that degeneration exists in kinetics and can be eliminated only by considering particle collisions and electromagnetic emissions.

In continuous velocity distributions, one can qualitatively and semi-quantitatively estimate the evaporating particle flux from a star (or the falling one to a star) under given boundary conditions. It gives the correct order of magnitude of mass and energy fluxes of the solar wind for the contemporary Sun. The solar wind speed at infinity is of the order of magnitude of the geometric mean of ion and electron thermal velocities in the solar corona at the exobase level. It means approximately forty times more intensive evaporation than that of neutral gas of the same temperature. The effect is the consequence of the plasma motion in the self-consistent potential electric field, which accelerates ions and decelerates electrons: no significant electric charge is gathered near the Sun. In this respect, the plasma has more “fluidity” than the neutral gas. The role of the magnetic field looks more complex. The Ampere ponderomotive forces can act both with gravity and the thermal pressure gradients and against these two. A “magnetic pump” can push the plasma in any direction depending on the “pump” structure and operation regimes.

The thermal expansion of the solar corona to the interplanetary space in the modern epoch seems the consequence of the evolution of the Sun (Веселовский 2008, 93-98). Today the Sun donates the matter to the local interstellar medium, but it accepted the matter from the local interstellar medium in previous stages of its evolution. The transformation between the regimes is almost unexplored. The answer to whether solar-like stars exist without the solar wind (or even with the accretion) can be found only from observations. We have no observational information yet of whether closest to the Sun solar-like stars in the local interstellar medium donates or accepted matter. Everything could depend on the initial conditions of given star evolution. At last, the intensity of the matter exchange between a star and its surroundings could be much less than mass flows circulating up and down in the star environment. In such a situation, stars with developed “turbo-sphere” could exist. In the solar case, its “turbo-sphere” is pressed by the gravitation to the inner corona, outside of which downward mass flows are much less than outward ones. The realization of the accretion regime or the expansion one depends not only on inner and outer boundary conditions but also on the temporal evolution of boundary conditions and initial conditions. The classical Parker model (Parker 1958, 644-675) with almost zero pressure at infinity does not answer how the external boundary conditions in the local interstellar medium could be changed to lead to the accretion on the Sun instead of the solar wind. At the outer boundary, where the dynamic pressure of the supersonic solar wind becomes comparable with the pressure of the local interstellar medium, a bow shock wave is formed. The values of solar wind velocity and density on different sides of the shock are connected via mass and momentum conservation conditions across the shock:

$$\rho^{-}M^{-} = \rho^{+}M^{+}, M^{+}M^{-} = 1$$

Indexes "–" and "+" denote upstream and downstream parameters. For a given pressure of the coronal base, the position of the shock is fixed by the pressure in the local interstellar medium. The distance between the shock and the Parker critical point decreases when the pressure in the local interstellar medium increases. The asymptotic pressure dependence on the upstream Mach number M^{-} is fixed by the mass conservation across the shock:

$$p_{\infty} = p_0 M^{-} \exp\left(M_*^2 - 2\upsilon - M^{-2} + \frac{1}{M^{-2}} \right) / 2$$

Here $\upsilon = GM_s/Rc^2$, M_* is the Mach number upstream of Parker's critical point (Velli 1994, L55-L58). What happens when the pressure difference between the coronal base and the local interstellar medium changes? Suppose the pressure in the local interstellar medium increases (or the pressure at the coronal base decreases). In that case, the shock shifts to the solar direction, and its amplitude decreases during this motion. Suppose the asymptotic pressure achieves the value of $p_{\infty}^{c} = p_0\exp(M_*^2/2-\upsilon)$. In that case, the shock achieves the Parker critical point heliocentric position $r_s = r_c = \upsilon/2$. The flow velocity discontinuity degenerates into the radial derivative of the flow velocity so that the shock disappears. The expansion becomes subsonic everywhere between the coronal base and the local interstellar medium. The subsequent increase of the pressure in the local interstellar medium leads to the decrease of the subsonic expansion velocity: the subsonic velocity increases with the heliocentric distance up to a certain distance but then gradually decreases to zero at the local interstellar medium boundary. The decrease of the expansion velocity leads to the decrease of the mass flux. The limiting regime of the subsonic expansion is the static corona. The subsequent increase of the pressure in the local interstellar medium initiates instabilities in the static corona and leads to its collapse. It could be understood from the following considerations. The increase of the external pressure compresses the static corona enhancing the electron concentration there. Radiant losses increase as the electron concentration square. These losses must be compensated by the thermal conductivity, which is proportional to the electron concentration logarithm only. The energy unbalance initiates oscillations in the static corona and its collapse in the form of the subsonic accretion: the corona falls to the solar surface with the acceleration due to the gravitation but brakes as approaching the solar surface because of the mass conservation. If the pressure in the local interstellar medium continues to increase, the accretion could become supersonic, and the bow shock could be formed near the solar surface. Let us imagine now that the external pressure in the local interstellar medium decreases (or the internal pressure at the coronal base increases). The decrease of the external pressure shifts the shock from the Sun. When the shock achieves the Parker critical point heliocentric position, it disappears, and the accretion transforms to the subsonic regime everywhere. The subsequent decrease of the external pressure leads to the accretion velocity decrease and then to the transformation from the subsonic accretion to the subsonic expansion via the static corona stage. The mass flux changes its direction to the opposite one. If the external pressure continues to decrease, the expansion transforms to the transonic solar wind. In such a way, the transformation from the expansion to the accretion and back (a kind of hysteresis) is realized via a catastrophe. Though an accurate analysis is needed to estimate the time duration of such a transition, one can suppose that it is approximately the time the sound spends to propagate from the coronal base to the Parker critical point heliocentric position (Korevaar 1989, 209-214). Seemingly, solar-like stars with accretion instead of the solar wind could exist. We do not know when and how the solar wind arose in the deep past and only guess that it could take place when the Sun came to the main sequence of its evolution and the nuclear fusion fired inside the Sun. We do not know when and how the solar wind will disappear in the future. We only guess that it would occur in case of the disappearance of the solar activity, which eventually maintains the hot solar corona, or when the Sun enters into the dense area of the local interstellar medium and/or into the powerful galactic magnetic field. Both scenarios are possible, but the conceivable time scale is the cosmogonic one.

The expanding solar wind flow creates the diamagnetic envelope around the Sun with the magnetic flux $\sim 5\times10^{-5}$ Gauss 4π (1 AU)^2 which is two or three orders of magnitude less than the magnetic flux in the local interstellar medium $\sim 5\times10^{-6}$ Gauss 4π (100 AU)^2. So plentiful a free energy supply in the external galactic magnetic flux allows considering the possible influence of the external magnetic flux on the solar wind electrodynamics. Field-aligned electric currents in the solar wind could connect the local interstellar medium to the solar wind. Electrons moving along the magnetic field ("up-streams") could serve as carriers of the information about the local interstellar medium. If one exploits the magnetosphere analogy, then solar polar coronal holes with the open magnetic configuration could be the Earth polar magnetic caps. According to the analogy, variations in its configuration follow external boundary conditions in the local interstellar medium. The large scale magnetic field in the diamagnetic envelope around the Sun is determined by electric currents in the Sun, interplanetary space, and local interstellar medium. The details of such a current system are poorly known, so we postulate the simplest one, neglecting the electric currents on the helio-pause and considering the constant external magnetic field in the local interstellar medium (Triskova and Veselovsky 1992, 297-300). The magnetospheric image of our "vacuum-like" model is the so-called "two dipole" model of the magnetosphere (Shabansky 1965, 125-147) (Антонова и Шабанский 1968, 801-811). The interplanetary magnetic field model is drawn in Fig. 5-1. The magnetic field is purely radial, constant, and oppositely directed in the northern and southern hemispheres on the solar surface. The magnetic field direction change is ensured by the surface electric current flowing in the equatorial plane of the interplanetary space. The Parker model describes the interplanetary field $\boldsymbol{B}_s$. This field smoothly transforms into the constant magnetic field $\boldsymbol{B}_0$ of the local interstellar medium at the external boundary.

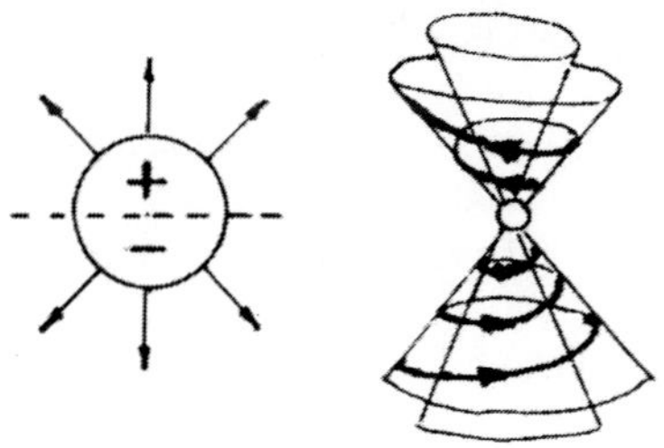

Fig. 5-1 The interplanetary magnetic field "vacuum-like" model

In spherical coordinates (r, θ, $\varnothing$), the model is as follows:

$$B_{rs} = B_s\left(\frac{r_0}{r}\right)^2 sign(\cos\gamma)$$

$$B_{\theta s} = 0 \qquad (5.1)$$

$$B_{\varnothing s} = -B_s\left(\frac{r_0}{r}\right)^2 \frac{\omega r}{V}\sin\theta sign(\cos\gamma)$$

$$\cos\gamma = \cos\theta\cos\alpha + \sin\theta\sin\alpha\cos(\varnothing_M - \varnothing),$$
$$\varnothing_M = \varnothing_0 + \omega(t - t_0)$$

$$\vec{B} = \vec{B}_s + \vec{B}_0$$

Here B_s = *const*, ω is the solar angular velocity, V is the solar wind velocity, α is the angle between the solar magnetic moment and the unit vector in θ-direction. Let us introduce the dimensionless distance based on the natural distance scale $L = r_0(B_s/B_0)^{1/2}$. The picture of magnetic field lines is shown in Fig. 5-2. The left drawing in Fig. 5-2 illustrates when the interstellar magnetic field is directed parallel to the solar magnetic moment. The central one – when the interstellar magnetic field is directed antiparallel to the solar magnetic moment. The right one – when the interstellar magnetic field is directed perpendicular to the solar magnetic moment. The solar rotation is neglected. Left and central drawings are meridional sections of the spatial field structure. The letter Z denotes zero points, numbers 1 and 2 – separatrixes, the dotted line – the sector boundary.

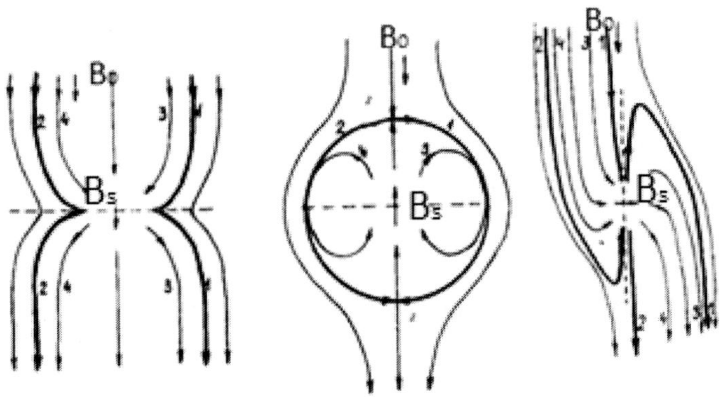

Fig. 5-2 Magnetic field line pictures

Magnetic field line equations in dimensionless cylindrical (z,ρ) or Cartesian (z,x) coordinates are as follows:

$$\frac{d\rho}{dz} = \rho\left[z + \left(\rho^2 + z^2\right)^{3/2} sign(z)\right]^{-1}$$ the left picture

$$\frac{d\rho}{dz} = \rho\left[z - \left(\rho^2 + z^2\right)^{3/2} sign(z)\right]^{-1} \text{ the central picture} \quad (5.2)$$

$$\frac{dx}{dz} = x\left[z + \left(x^2 + z^2\right)^{3/2} sign(x)\right]^{-1} \text{ the right picture}$$

The analytical solution of Eqs. (5.2) leads to the magnetic field line equation in the form:

$$x^2 + y^2 = 2x\sqrt{x^2 + y^2}\frac{L^2}{y^2} + C\frac{x^2 + y^2}{y^2}$$

Here C is the integration constant. The picture of magnetic field lines when the solar magnetic moment and the interstellar magnetic field are directed perpendicular to the solar rotation axis is shown in Fig. 5-3. Zero points are possible in this case. The choice between an open magnetic configuration and a closed one depends on the interstellar field component parallel to the solar magnetic moment – this component shifts the magnetic field lines, as shown in the right picture in Fig. 5-2.

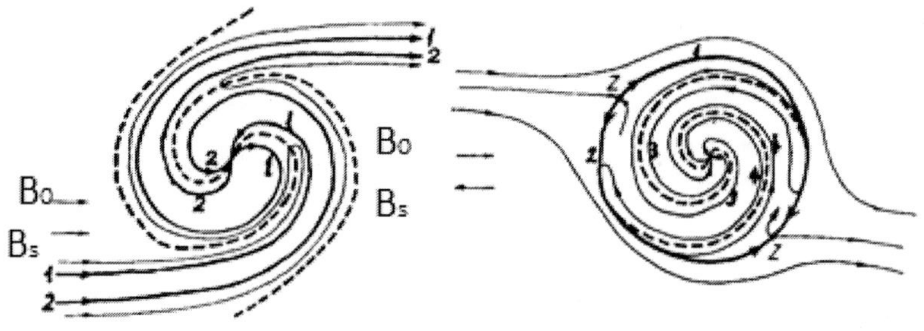

Fig. 5-3 The picture of magnetic field lines when the solar magnetic moment and the interstellar magnetic field are directed perpendicular to the solar rotation axis

The solar magnetic moment could be directed perpendicular to the solar rotation axis in a solar activity maximum. One can see that the magnetic field configuration in the diamagnetic envelope around the Sun could be alternately open and closed in a solar activity maximum. In the plane of the equatorial section ($z = 0$, $z||\omega$, $x||\boldsymbol{B}_0$, $\cos\gamma = \cos(\varphi_M\text{-}\varphi)$), the magnetic field line equation in dimensionless variables is as follows:

$$\frac{dy}{dx} = \left(y + \frac{\omega L}{V} x\sqrt{x^2 + y^2} \right) sign(\cos\gamma)$$

$$\left[\left(x + \frac{\omega L}{V} y\sqrt{x^2 + y^2} \right) sign(\cos\gamma) + \left(x^2 + y^2\right)^{3/2}\right]^{-1}$$

The interstellar magnetic field direction is unknown. Sometimes, they assume that it is parallel to the interstellar gas velocity (i.e. close to the ecliptic). In a solar activity minimum, when the solar magnetic moment is directed along the solar rotation axis (see Fig. 5-1) and the interstellar field is directed perpendicular to this axis, the picture of magnetic field lines is shown in Fig. 5-4.

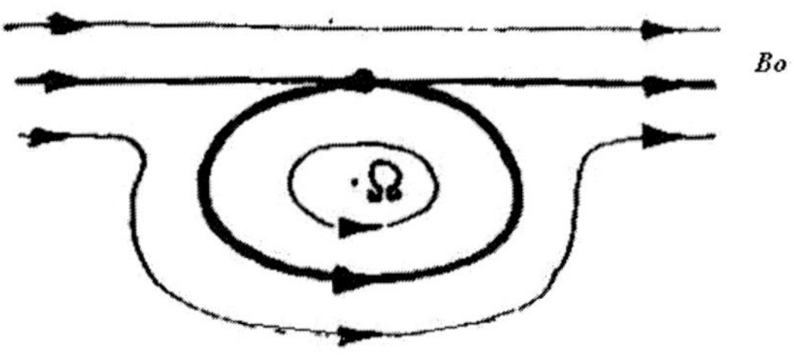

Fig. 5-4 The picture of magnetic field lines in a solar activity minimum

In this case, the magnetic field in the diamagnetic envelope around the Sun is the sum of two addendums $\boldsymbol{B}_{\varphi s}$ and $\boldsymbol{B}_0$ in Eq. (5.1). The magnetic field line equations in dimensionless variables are as follows:

$$\frac{dx}{dz} = ay\left[ax + \left(x^2 + y^2 + z^2\right)sign(z)\right]^{-1}$$

$$\frac{dx}{dy} = \frac{\omega L}{V} y\left[-\frac{\omega L}{V}x + \left(x^2 + y^2 + z^2\right)sign(z)\right]^{-1}$$

Here $z||\boldsymbol{\omega}$, $y||\boldsymbol{B}_0$; $\boldsymbol{B}_s$ is situated in the plane xy. Let us note that previous considerations, which illustrate the possibility of the diamagnetic envelope with an open magnetic configuration at certain time moments, neglect the shielding of the diamagnetic envelope from the local interstellar medium magnetic field penetration (Veselovsky 1982, 209-213). The shielding could be created by electric currents on the helio-pause, neglected in the above consideration.

We do not know how the solar wind originates close to the Sun and how it dissolves in the local interstellar medium. The solar wind velocity at the Earth orbit can variate in the broad interval, but one never observes values less than 200 km s^{-1} and more than 3000 km s^{-1}. Nobody knows why these limits exist and why the average solar wind density is several particles per cubic centimetre and not several thousands of particles per cubic centimetre or, conversely, a few particles per cubic metre. The evolution of these limits remains enigmatic.

Sources and tanks of solar wind matter

The imagination about homogeneous sources of the solar wind is approximate. It describes the phenomenon very schematically when common properties (like the spatial scale, the lifetime, the character density, the velocity, the temperature, the velocity distributions, the power and the spectrum of electromagnetic emissions) are the subject of interest. The concept of tanks of solar wind matter is imaginative too because the Sun has neither rigid boundaries nor tanks with holes for the matter exhaust. The solar matter confinement is realized mainly due to the gravitation and the magnetic field. The role of the gravitation confinement decreases while the role of the magnetic one increases when the heliocentric distance increases. However, the magnetic field inhibition of the transverse plasma motion is valid only in the absence of electric fields. There is no such prohibition in the general case, and transverse plasma motions are observed in the solar corona.

In hydrodynamics, a source means the place of the flow exit from under the surface. As applied to the solar wind, a source means a part of the solar surface emanating the plasma into the interplanetary space.

Because there are no rigid borders on the Sun, the concept of "source" is conditional. The main problem in tracing a control element of the solar wind from its source to the point of observation in the interplanetary space is created by the lack of reliable physical techniques of solar wind velocity measurements. The moving element of the solar wind should be reliably separated from the background to trace its motion and reconstruct the line of flow. Suppose one traces the elements of the solar wind. In that case, the nonstationary topologically multiply connected three-dimensional picture of flow with numerous streams, "islands", bifurcation points, branches, junctions, and whirls around stagnation points in the concomitant reference frame are manifested. Self-consistent magnetic fields partly control the motion.

Does the solar wind blow from the whole solar surface or its specific part? The answer depends on the chosen surface around the Sun, its position and duration of existence (Веселовский 2017, 269-272). For simplicity, let us consider an instantaneous flow picture across concentric spheres around the Sun. All available observations testify to the impossibility of the plasma confinement neither by the magnetic field nor by the gravitation so that it leaves into the interplanetary space forever. The imaginary boundary (the so-called "turbo-pause") exists. Below the "turbo-pause", the plasma slowly ascends, moving up and down many times. After the “turbo-pause” passage, the return to the Sun is impossible: the almost radial flow originates, which is called the solar wind, and the velocity direction becomes outward finally and unalterably. In reality, the “turbo-pause” is not obligatory continuous and simply connected. Its details are poorly known. Simply connected single solar sources meet not so often as multiply connected and complex ones due to the influence of coronal magnetic fields. Flows from multiply connected sources sometimes join into a united, approximately homogeneous flow like pseudo streamers above active solar regions. Opposite processes of the initially homogeneous flow splitting also take place. The image arises of the transition from turbulent up and down motions near the Sun to the more organized fast radial outward flow far from the Sun with the turbulence in it. It is important to note that ion motions back to the Sun were never observed at heliocentric distances further than six solar radii: the almost radial subalfvenic solar wind presents there continuously and everywhere.

No stationary sources on the Sun exist – the lifetime of all such sources is limited. Besides co-rotating high-speed streams living up to several solar rotations, there are many transient phenomena in the solar wind generated by coronal mass ejections with lifetimes from days to several hours. These phenomena are moving from the Sun clouds, loops, and envelopes of various shapes. There are two natural time scales in the solar wind: the solar rotation period and the time of flight of the plasma between the Sun and the Earth with the average solar wind velocity ~ 400 km s^{-1}. The Struhal number $Str = vt/l$ (where v is the character velocity, t is the character time variation scale in the source, l is the character length) can easy segregate quasi-stationary ($Str \ll 1$) situations from transient ($Str \gg 1$) ones. Intermediate cases ($Str \sim 1$) are poorly studied: one of the examples is transient coronal holes. The division based on the Struhal number is not invariant relative to the moving reference frame because the character velocity is included in the Struhal number definition. Therefore, the same phenomenon can be treated as quasi-stationary or transient depending on the frame of reference. The same remark is valid for the division between laminar and turbulent flows based on the Reynolds number: in the concomitant reference frame, a flow can look as laminar, but in the spacecraft reference frame, it seems short-lived and turbulent.

Coronal streamers, pseudo-streamers, open, closed, and intermittent magnetic structures in the solar corona are considered solar wind sources. All these are situated at heliocentric distances between 0.1 and 6 solar radii. It is challenging to visualize solar wind sources at heliocentric distances smaller than 0.1 solar radii because of the magneto-hydrodynamic turbulence. The upward average flow at such distances is much weaker than stochastically directed, much more powerful turbulent flows. It is seen in spicules and X-ray jets considered as potential suppliers of matter in the solar wind. In spicules, the chromosphere material with temperatures ~ 10^4 °K is elevated and heated up to ~ 10^5 °K in the transition region. Then it cools due to the radiant emission and the thermal conductivity and falls to the chromosphere. It is difficult to estimate what part of the material is dispersed and lost in the corona with the temperature ~ 10^6 °K and what part of this part comes to the solar wind because no standard technique to observe so different plasma states exists. If the matter disbalance between up and down convective motions in spicules existed, and if the disbalance were of the solar wind mass flux order, then spicules would be the matter suppliers for the solar wind. Seemingly, observations contradict this hypothesis, and a considerable portion of matter comes to the corona and the solar wind from the inter-spicule space. In turn, it comes from photosphere pores that occupy only 10^{-6} parts of the photosphere surface. The photosphere granulation (and the thermal convection in the form of chromosphere spicules) “knows” nothing about the solar wind. Solar wind mass and energy fluxes are much less than ones circulating in the photosphere and the chromosphere. Possibly, the whole solar surface takes part in producing the solar wind mass flux. Many already completed, current, and planned space missions (“SOHO”,

"Wind", "ACE", "Stereo", "SDO", "Solar orbiter", "Parker solar probe") declare the exploration of the relationships between the solar wind, the corona, and lower solar atmosphere layers as its primary goal.

External relative to the Sun sources of the solar wind matter supply also exist but are sidetrack. However, very far from the Sun, the neutral interstellar gas brings to the foreground. Neutral atoms moving to the Sun are ionized by the charge exchange and the solar ultraviolet emission. Newly born ions are picked up by the solar wind initiating overloading and decelerating the solar wind flow. Comets and interstellar dust particles with trajectories situated in the vicinity of the Sun also bring some deposit to the solar wind matter supply. The heliocentric distance where the solar wind dynamic pressure is close to the pressure in the local interstellar medium is estimated as ~ 100 AU. No stationary boundaries are expected because of ample supplies of free energy in the local interstellar medium. Known is no observational information about the solar wind mixing with the interstellar medium. A buffer zone ought to exist where such mixing occurs: the solar wind in separate streams penetrates the local interstellar medium, intermixes with the interstellar material. Eventually, it resolves in the interstellar reservoir containing much more matter of the interstellar wind. The spatial character scale of the local interstellar medium is of the order of ~ 10^{18}-10^{19} cm, which corresponds to the distance between the

Sun and closest stars (a few light-years). For the interstellar gas concentration of ~ 1 cm^{-3}, the total number of particles in the local interstellar medium is comparable with the total number of particles in the Sun. The renovation of this number due to the interstellar wind blowing with the velocity of a few tens of kilometres per second spends tens of thousands of years. The deposit of the solar wind in this matter balance is negligible: the solar wind enters into the local interstellar medium as a small admixture. It dissolves there (due to the turbulent diffusion) at much shorter distances than interstellar ones. Even an approximate estimate of the distance and time for the solar wind to dissolve looks uncertain because of the instability typical for the problem of flow entrainment from a weak source with a more powerful background flow.

The additional information about the solar wind matter sources can be extracted from the chemical composition of the solar wind (Geiss et al. 1970, 458-483). There are various ions in the solar wind. The study of its masses and charges allows its use as test particles (because of low concentrations of specific ions) in tracing solar spatial structures in the interplanetary space and exploring the chemical composition of the Sun and its atmosphere. The solar wind ion composition formation is too complex; we will consider involved processes only qualitatively. Many comparable factors (electromagnetic, gravitational, molecular kinetic, inertial, and atomic) control these processes, and it stipulates the whole picture's diversity. There are a lot of variants. Besides potential and induction electric fields, presented are static and variable magnetic fields created by electric currents in the solar wind. Electric fields separate ions by their electric charges; in dynamical situations – according to the charge to mass ratio. Magnetic fields create the Lorentz force, which linearly depends on the nonrelativistic ion's velocity and separate ions by its velocities. The electrostatic separation of ions from electrons is impeded because of the plasma quasi-neutrality though the Debye radius linearly increases with the heliocentric distance. The gravitation keeps ions closer to the Sun than electrons. It creates the polarization electric field with the potential of the order of the electron thermal energy, i.e. hundred electron-volts in the solar wind acceleration region. The combined effect of the gravitation and the electric field forms populations of trapped and passing particles, but the magnetic field intensely transforms the picture. Inertial forces are essential in dynamic nonstationary processes and can separate ions according to their masses and accelerations when the free plasma spread occurs. Cross-sections of collisions depend on the particle's charge, velocity, and other parameters. There are elastic and inelastic collisions, ionizations, recombinations, excitations in partly ionized plasma in the chromosphere where the solar wind ion composition is formed. In almost equal ion and electron temperatures, the ionization state is determined by the elastic ionization and di-electron and radiative recombinations. It explains the separation of atoms in the solar wind according to their first ionization potential. It depends on whether the separation mechanisms have time to act during the flight time of the matter passage through the region. Neutral atoms are not involved in solar wind formation in regions of rare collisions. Therefore, the depletion of ions with the high first ionization potential is registered in the fast solar wind compared to the slow one. The difference is estimated as 1.5 – 2. The slow wind is closer to the local thermodynamic equilibrium, but it is inhomogeneous and enriched with low ionization potential ions than its photosphere abundances. Let us note that the distinction between "fast" and "slow" wind flows is somewhat conditional: no contradiction exists in the description of the solar wind phenomenon as many individual solitons moving from the Sun and superimposed each other. These solitons are challenging to discriminate, especially at large heliocentric distances where the mixing leads to the formation of the solar wind as a whole.

The so-called "ionization freezing" means that ionization and recombination times are much longer than the motion flight time: the fast prominence with temperatures ~ 10^4 °K passing through the corona with temperatures ~ 10^6 °K has no time to be fully ionized. It is why the singly ionized helium has no time to transform to the α-particle and is registered in coronal mass ejections on spacecraft at the Earth orbit from time to time. Observations show that ions Fe^{+10} and Fe^{+9} prevail in high-speed streams in the ecliptic and correspond to the "ionization freezing" temperature ~ 1.1×10^6 °K in the equatorial coronal base. In contrast, ions Fe^{+11} and Fe^{+10} prevail in the fast wind in high heliolatitudes and correspond to the higher "ionization freezing" temperature in the polar coronal base. In the slow solar wind, particles collide more often and the flight time of its motion is longer than in the fast solar wind. Quite often, the correspondence between the ion mass and the ion temperature takes place. It could be observed in several situations, for example, in the ion heating by Alfven and ion cyclotron waves. Both wave types lead to the enhancement of transverse ion temperature components. In the case of the falling wave spectrum, heavier ions become hotter. Another possibility is the transfer of the kinetic energy of different particle groups into the heat in mutually penetrating streams when the thermal dispersion is proportional to the ion mass. First of all, this situation relates to the enhancement of the field-aligned ion temperature component.

The solar wind chemical composition corresponds, in general, to the chemical composition of the Sun. However, prudence is needed because the equilibrium between ionization and recombination in the solar atmosphere is strongly disturbed by supra-thermal particles and hard electromagnetic radiations. On the other hand, in-situ measurements on spacecraft show that the solar wind chemical composition strongly fluctuates. It is difficult to compare them with the indirect integral data derived from metallic foils exposed in the solar wind and extracted from meteorites, comet, and planetary samples. Gathered were data for ratios D/H, $3He^3/4He^4$, $(D+3He^3)/H$, Ne^{20}/Ne^{22}, and Ne^{21}/Ne^{22} obtained in experiments with aluminium foils exposed on the Moon (Geiss and Gloecker 2003, 3-18) in the framework of the “Apollo” program and with mica plates – on “Soyuz”, “Zond 8”, and “Luna 16” spacecraft (Boltenkov et al. 1972, 1487-1491). These were interpreted as the proto-solar matter chemical composition because the isotopic fractionation in the fast solar wind is much weaker than the one registered for energetic particles and variates in the range of 1–2% on an atomic mass unit for N, Ne, Mg, Si, Ar, Fe (Kallenbach 2003, 305-316). However, the isotopic composition of lithium and rare gases in the solar wind and the terrestrial matter differs significantly; for D, Li, Be, B, the difference could be connected to the Sun’s nuclear processes. However, such extrapolations into the deep past are very uncertain, and so do cosmogonic scenarios based on the extrapolations. Though abundances of heavy multivalent ions in the solar wind correspond, in general, to the ion composition and the ionization temperature of the solar corona, rare registrations of low and singly ionized atoms directly point to its non-coronal origin. Such ions could indicate the chromospheric material or trapped pick-up ions created by the charge exchange and the photoionization from comets, meteorites, planets etc.

We cannot satisfactorily explain why the helium abundance in the solar wind at the Earth orbit is 25–30% relative to protons and why more rich abundances of heavy ions have never been observed despite many separation mechanisms in the solar wind.

Solar wind acceleration region energetics

Until now, there is no answer to the fundamental question: how the "hot" corona heated up to ~ 10^6 °K can appear above the "cold" photosphere heated only to ~ 6×10^3 °K without the violation of the energy degradation law. The heat source is situated in the solar core, where the nuclear fusion takes place, and the photosphere is situated closer to the core than the corona.

Estimates show that the energy flux needed to maintain the corona is one hundred times weaker than the flux entering the chromosphere from the photosphere. Only a tiny part of this flux is spent as the solar wind. Let us note that directions of mass and energy fluxes do not obligatorily coincide in the solar wind. These two fluxes coincide in the supersonic superalfvenic wind in the interplanetary space but very often have different directions in the solar corona. There is no universal heating mechanism in the corona. The free magnetic energy in electric currents and propagating electromagnetic and magneto-hydrodynamic waves of various types look promising candidates to serve as the main power supply for coronal heating. In the framework of such a paradigm, electromagnetic ponderomotive forces push up the coronal plasma from the Sun, complementing the thermal evaporation of hot particles into the interplanetary space.

Let us note that as early as in 1940th the hypothesis was formulated that the acoustic noise generated in the solar convective zone could heat the corona (Alfven 1947, 211-219) (Biermann 1948, 161-177) (Schwarzschield 1948, 1-5) (Schatzman 1949, 203-228). It was based on the statement that the coefficient of acoustic noise generation in the volume unit strongly depended on the velocity, and the maximal velocities took place in the convection zone. Therefore, most acoustic noise should be generated in the thin (the thickness ~ 100 km) spherical layer centered in the center of the Sun and situated inside the solar convective zone. The solar magnetic field specifies the noise as created by Alfven and magneto-sonic waves. Alfven waves propagate along the magnetic field while fast and slow magneto-sonic waves – perpendicular to the magnetic field. However, fast magneto-sonic waves cannot penetrate the corona from the photosphere (Kuperus 1969, 713-739) (Osterbrock 1959, 347-388); therefore, the wave energy flux is carried mainly by Alfven and slow magneto-sonic waves. Because radio-sounding revealed oscillations of Faraday's rotation of the polarization plane with periods ~ 5 minutes, one assumed that Alfven waves prevailed in the corona (Ефимов и др. 1993, 143-152). The Alfven wave shakes the plasma element forward and backwards (due to the Ampere force created by the wave). This shaking heats the element and, on average, accelerates it in the direction of the wave propagation. The Ampere force induced by the wave is as follows:

$$F_a = \frac{1}{4\pi}\left[\vec{j} \times \vec{B}\right]_r - \rho\left[\left(\vec{v} \cdot \nabla\right)\vec{v}\right]_r \tag{5.3}$$

Here ***j***, ***B***, ***v*** are the electric current, the magnetic field, and the plasma velocity in the Alfven wave, correspondingly. In the case of the spherical symmetry and the external magnetic field $\boldsymbol{B}_0$ described by Eq. (4.5), Washimi (Washimi 1973, 1373-1383) reduces Eq. (5.3) to the form:

$$F_a = \frac{1}{8\pi}\left(\frac{\partial}{\partial r} + \frac{2}{V_{gr}}\frac{\partial}{\partial t} + 2\kappa\right)|B|^2 \tag{5.4}$$

Here $V_{gr} = V_{SW} + V_A$ is the group velocity of the wave, V_{SW} is the solar wind velocity, $V_A = B_0/(4\pi\rho)^{1/2}$ is the Alfven speed, κ is the Alfven wave absorption factor. Eq. (5.4) corresponds with the more general expression for the ponderomotive force calculated from thermodynamics (Washimi and Karpman 1976, 528-531). A spherically symmetric solar wind model with the magnetic field specified by the Eq. (4.5) and additional energy-momentum sources created by Alfven waves was elaborated by Chashei and Shishov (Чашей и Шишов 1987, 705-711) (Чашей и Шишов 1988, 157-166) as follows:

$$B(r) = B_0\left(\frac{R}{r}\right)^2 \tag{5.5}$$

$$\frac{d}{dr}\left(\rho V r^2\right) = 0 \tag{5.6}$$

$$\rho V\frac{dV}{dr} + \frac{dP}{dr} + \rho g(r) + F_a = 0 \tag{5.7}$$

$$\frac{1}{r^2}\frac{d}{dr}r^2\left[\rho V\left(\frac{V^2}{2} + \frac{5V_T^2}{2} - \frac{V_g^2}{2}\right) - \chi\frac{dT}{dr} + H_r^a\right] = -Q_{em} \tag{5.8}$$

$$P = \rho V_T^2 \tag{5.9}$$

Here $g_s(r) = g_{s0}(R/r)^2$ is the acceleration of gravity, $V_g = V_{g0}(R/r)^2$ is the speed at liftoff, $V_T = (2kT/m_p)^{1/2}$ is the isothermal sonic speed, $\chi = [7\times10^{-7}$ erg/(cm s $^{o}K^{7/2})]T^{5/2}$ is the classic coefficient of the electron thermal conductivity, Q_{em} are radiation losses:

$$Q_{em} = n^2 L(T) \tag{5.10}$$

The function $L(T)$ is tabulated (Cox and Tucker 1969, 1157-1167), F_a and H_r^a are radial components of the wave pressure force and the Alfven wave energy flux specifying by the equations (Jacques 1977, 942-951):

$$\frac{1}{r^2}\frac{d}{dr}\left[r^2 (V + V_A) W_\omega^a\right] = -\gamma_\omega^a W_\omega^a + \frac{V}{2}\frac{dW_\omega^a}{dr} \tag{5.11}$$

$$H_\omega^a = (V + V_A) W_\omega^a, H_r^a = \int H_\omega^a d\omega,$$

$$W^a = \int W_\omega^a d\omega, F_a = \frac{1}{2}\frac{dW^a}{dr} \tag{5.12}$$

Here W_ω^a is the Alfven wave energy flux spectral density, γ_ω^a – absorption coefficients. Eqs. (5.5)-(5.12) were complemented by the boundary conditions as follows:

$$T(r) \to 0 \text{ under } r \to \infty \tag{5.13}$$

$$T(r) \cong 10^{4\,o} K \text{ under } r = r_* \cong R \tag{5.14}$$

$$\frac{dT}{dr} = 0 \text{ under } r = r_* \tag{5.15}$$

$$W_\omega^a = W_\omega^a (r_*) \text{ under } r = r_* \tag{5.16}$$

$$V^2 = V_T^2 + u_2^2, r^2 \frac{dV_T^2}{dr} + g_0 \left(\frac{r_*}{r}\right)^2 = \quad \text{under } r=r_c \qquad (5.17)$$

$$-\frac{1}{\rho}\left(F_a - u_2^2 \frac{d\rho}{dr}\right)$$

Eq. (5.14) is defined by the form of the function $L(T)$ (Eq. (5.10)) which allows significant radiation losses only when $T(r) \geq 10^4$ °K so that one can fix this temperature value to be the temperature of the transition layer ($r = r_*$) separating the chromosphere from the corona. Eq. (5.15) means that the thermal conductivity flux from the corona to the chromosphere is absorbed in the transition layer. The position of the transition layer ($r = r_*$) is sliding and is adjusted during the solution procedure. The Alfven wave energy flux spectral density $W_\omega{}^a(r_*)$ is specified as follows:

$$W_\omega^a(r_*) = \begin{cases} A\omega_0^{-1}, \omega < \omega_0 \\ A\omega_0^{-1}\left(\dfrac{\omega_0}{\omega}\right)^n, \omega > \omega_0 \end{cases} \qquad (5.18)$$

Here $\omega_0 \approx 2\ 10^{-2}$ s^{-1} is the energy-containing frequency that corresponds to known five-minute oscillations. The following conditions define the normalizing constant:

$$A = \eta_{0*}(n-1)\frac{W^a(r_0)}{n}, W^a(r_0) = \frac{B_0^2}{8\pi}, \eta_{0*} = \frac{4\sqrt{\rho_0 \rho_*}}{\sqrt{\rho_0} + \sqrt{\rho_*}} \qquad (5.19)$$

Here $W^a(r_0)$ is the total wave energy at a specific heliocentric distance $r_0 > r_*$. The condition $\beta_0 = 4\pi P_0/B_0^2 = 1$ determines the distance r_0. The coefficient η_{0*} describes the penetration between r_* and r_0; an index n could be chosen 5/3 or 3/2. Eq. (5.17) is the demand of the continuous transition through the modified sonic speed at the critical point $r = r_c$. The value u_2 is defined as follows:

$$u_2^2 = \frac{(3V + V_A)W^a}{4(V + V_A)\rho} \tag{5.20}$$

The heliocentric distance of the critical point is determined by equalizing the Alfven wave pressure gradient to the gravitational force. Absorption coefficients $\gamma_\omega{}^a$ are given. Eqs. (5.5)-(5.20) contain the only free parameter – the magnetic field B_0 at the coronal base.

Calculations (according to Eqs. (5.5)-(5.20)) show that Alfven waves propagate in the linear regime close to the transition layer. The waves with frequencies $\omega > \omega_0$ are damped linearly. The coronal heating due to the wave passage demonstrates its maximum at the frequency ω_0 of the maximal damping. The coronal temperature maximum is situated at the heliocentric distance where the maximum of damping occurs. The corona is transparent for waves with frequencies $\omega < \omega_0$, and the relative amplitude of such waves increases with the heliocentric distance. When the relative amplitude approaches the unit, Alfven waves can generate fast magneto-sonic waves (due to nonlinear instabilities not included to Eqs. (5.5)-(5.20) directly), which heat the corona more effective. The wave pressure increase leads to the additional acceleration of the solar wind in the heliocentric distance interval between 10 and 20 solar radii. Depending on the value of B_0, one can identify the two regimes, specifically, the weak-field regime $B_0 < B_{sat} \approx 2$ Gauss and the saturation regime $B_0 > B_{sat}$. In the weak-field regime, thermal and magnetic pressures are close. The Alfven wave energy flux is formed at the transition region base r_*. The increase of B_0 leads to a lower transition region base, increasing the heliocentric distance of the maximum temperature ($T_{max} \propto B_0^{4/5}$). The solar wind mass flux increases while the solar wind terminal velocity decreases. The minimum terminal velocity $V_\infty = 350$ km s^{-1} occurs for $B_0 = B_{sat}$ when the temperature maximum $T_{max} \approx 2\times10^6$ °K achieves at the heliocentric distance ~ 1.5 R. In the saturation regime, the magnetic pressure is much more than the thermal one at the coronal base. When B_0 increases, the solar wind mass flux remains almost constant while the solar wind energy flux increases. This leads to the solar wind terminal velocity increase from $V_\infty = 350$ km s^{-1} for $B_0 = B_{sat}$ up to $V_\infty \cong 1000$ km s^{-1} for $B_0 >> B_{sat}$. The deposit of Alfven waves to the energy balance of the corona (and the solar wind) dominates in the saturation regime.

The role of the solar radiative emission remains still unclear. It can be described by introducing the gradient of the pressure of light into the momentum conservation equation. The pressure of light could push the plasma from the Sun. At the same time, every emitted photon creates the recoil momentum so that a certain recoil momentum pressure will act at the photosphere transmission cutoff surface, which will press the plasma to the Sun. The problem is well known in laser-induced nuclear fusion. One usually considers the plasma (where light propagates) either optically thin or optically thick. The first choice implies that the solar light propagation and the plasma motion are uncoupled. The second one needs to take into account the complex energy redistribution between the plasma and the light. There is the commonly accepted point of view that solar light does not influence the solar wind. The solar wind mass flux is often established as an additional external condition in solar wind models. The fact that the flux is of the order of the total rest mass of the solar material taken away with the solar light is interpreted as a random coincidence. The current rate of solar mass losses due to the solar wind is estimated as 2×10^{18} g s^{-1} (of the order of 10^{-14} of the solar mass per year). It is negligible in the total solar mass balance on the contemporary evolution stage of the Sun as a star. For the solar mass of about 2×10^{33} g and the assumed age of about 10^{10} years, the main deposit to the solar rest mass loss is contributed by the solar light and not by the solar wind. However, caution is needed to extrapolate such estimates into the deep past. The previous evolution of a system determines which stationary state will be realized – the solar corona is not an exception. In this connection, besides the wave energy source situated inside the Sun, theory, in general, does not eliminate that energy sources existed in the past outside the Sun. For example, Hester, with colleagues (Hester et al. 2004, 1116-1117), supposes that the Sun, along with some other stars with relatively small masses, were formed in surroundings of a giant star as a result of the hot gas ejection from this star accompanied by the shock wave passage through its atmosphere. After the Sun had been formed, its atmosphere, heated by the radiation of the large parent star, was an evaporating spherical layer. According to the authors, such processes are observed, for example, in the Trifid Nebula. The evolution of the parent star terminated with its explosion, which enriched the nearest vicinity of the Universe in the chemical elements, including ^{60}Fe, synthesized during the explosion. As proof of this scenario, the authors indicate the presence of the ^{60}Ni isotope in pieces of meteorites found on the Earth. This isotope could be produced only by the radioactive decay of ^{60}Fe. Within the framework of this scenario, which assumes the former existence of an external source of the heating of the solar atmosphere, the transonic character of the expansion of the solar corona could be a relic of the evolutionary stage of the Sun when it was near its parent star. A model of the relic spherically symmetric solar wind (preserved since the moment when the external source of the coronal heating was switched off) with the magnetic field given by Eq. (4.5) is specified by mass, momentum, and energy conservation equations (Pisanko 2005, 426-430):

$$nVr^2 = C(215R)^2 \tag{5.21}$$

$$nm_pV\frac{dV}{dr} = -2\kappa\frac{dnT}{dr} - nm_p\frac{GM_s}{r^2} \tag{5.22}$$

$$\begin{aligned}\frac{1}{r^2}\frac{d}{dr}\left(k_0T^{5/2}r^2\frac{dT}{dr}\right) - k_1n^2\sqrt{T} - k_2\frac{n^2}{\sqrt{T}} = \\ -2kTV\frac{dn}{dr} + 3knV\frac{dT}{dr}\end{aligned} \tag{5.23}$$

Here $k_1 = 1.5\times10^{-27}$ erg cm^{-2} s^{-1} ^{o}K$^{-1/2}$, $k_2 = 6\times10^{-22}$ erg cm^{-2} s^{-1} ^{o}K$^{1/2}$, C is the solar wind mass flux at the Earth orbit. Due to the high coronal temperature and the high ionization degree, most of the coronal electromagnetic emission falls at X-rays, (and the far-ultraviolet) and this radiation is one of the leading heat loss mechanisms of the solar corona (Гибсон, 1977, 367). At coronal temperatures of ~ 10^6 ^{o}K, the thermal bremsstrahlung radiation and the recombination radiation are essential. The second addendum in the left-hand side of Eq. (5.23) describes the energy loss due to the thermal bremsstrahlung radiation; the third one describes losses due to the recombination radiation (Арцимович, 1966, 87-89). The top of the photosphere, characterized by the temperature minimum, is chosen as the inner boundary:

$$\left.\frac{dT}{dr}\right|_{r=R} = 0$$

$$T\big|_{r=R} = T_0$$

$$n\big|_{r=R} = n_0 \qquad (5.24)$$

Here T_0 and n_0 are the temperature and the plasma density in the minimum temperature region, respectively. Let us introduce dimensionless variables, using the scale height in the region of the temperature minimum $H = kT_0R^2(m_pGM)^{-1}$ as the unit of the heliocentric distance, n_0 as the density unit, T_0 as the temperature unit, and $(2kT_0/m_p)^{1/2}$ as the unit of the velocity. Introducing a new independent variable $y = 1 - \alpha r^{-1}$, where $\alpha = RH^{-1}$, so that $y = 0$ for $r = \alpha$ $y = 1$ for $r = \infty$, and eliminating the density with the help of Eq. (5.21), we can perform algebraic transformations to obtain:

$$\frac{dV^2}{dy} = \frac{V^2}{V^2 - T}\left(\frac{4T}{1-y} - 2\frac{dT}{dy} - \alpha\right) \tag{5.25}$$

$$\frac{d^2T}{dy^2} = \beta\left(1 + \frac{\varepsilon}{T}\right)\left(\frac{\delta}{\alpha T V}\right)^2 + \frac{\gamma\delta}{\alpha T^{3/2}}\left(\frac{4}{1-y} + \frac{1}{V^2}\frac{dV^2}{dy} + \frac{3}{T}\frac{dT}{dy}\right) - \frac{5}{2T}\left(\frac{dT}{dy}\right)^2 \tag{5.26}$$

$$T\big|_{y=0} = 1 \tag{5.27}$$

$$\left.\frac{dT}{dy}\right|_{y=0} = 0 \tag{5.28}$$

$$V\big|_{y=0} = \frac{\delta}{\alpha^2} \tag{5.29}$$

Here $\beta = n_0^2H^2k_1T_0^{-3}k_0^{-1}$, $\gamma = n_0H\kappa^{3/2}2^{1/2}m_p^{-1/2}T_0^{-2}k_0^{-1}$, $\delta = C(215\alpha)^2m_p^{1/2}n_0^{-1}(2\kappa T_0)^{-1/2}H^{-2}$, $\varepsilon = k_2(k_1T_0)^{-1}$ are dimensionless parameters. The boundary condition for the velocity (5.29) was obtained from Eq. (5.21) with the allowance for the boundary condition for the density (5.24). Eqs. (5.25)-(5.29) have a singular (critical) point at $V^2 = T$. To ensure the boundedness of the first derivative of the squared velocity (and of the second derivative of the temperature, respectively) at the singular point, the following condition should be valid:

$$\frac{4T}{1-y} - 2\frac{dT}{dy} - \alpha = 0 \tag{5.30}$$

Its integration gives the temperature asymptotic near the singular point:

$$\begin{aligned} T(y) = T_1\left(\frac{1-y_1}{1-y}\right)^2 + \frac{\alpha}{2}(y_1 - y) \\ \frac{1-y_1-y+(1/3)(y_1^2+yy_1+y^2)}{(1-y)^2} \end{aligned} \tag{5.31}$$

Here $T_1 = T(y_1)$ is the temperature in the closest vicinity of the singular point. The value dV^2/dy at the singular point is estimated from the L'Hospital rule:

$$\frac{dV^2}{dy} = T\frac{\dfrac{4}{1-y}\dfrac{dT}{dy} + \dfrac{4T}{(1-y)^2} - 2\dfrac{d^2T}{dy^2}}{\dfrac{dV^2}{dy} - \dfrac{dT}{dy}}$$

Therefore, taking into account that according to Eq. (5.30)

$$\frac{dT}{dy} = \frac{\alpha}{2} - \frac{2T}{1-y}$$

at the singular point, and, hence,

$$\frac{d^2T}{dy^2} = \frac{6T}{(1-y)^2} - \frac{\alpha}{1-y}$$

we have the following quadratic equation for the determination of $U = dV^2/dy$ at the singular point:

$$U^2 - \left(\frac{\alpha}{2} - \frac{2T}{1-y}\right)U - \frac{4}{1-y}\left(\alpha - \frac{4T}{1-y}\right) = 0$$

It has the solution

$$U \equiv \frac{dV^2}{dy} = \frac{\alpha}{4} - \frac{T}{1-y} + \sqrt{\left(\frac{\alpha}{4} - \frac{T}{1-y}\right)^2 + \frac{4}{1-y}\left(\alpha - \frac{4T}{1-y}\right)}$$

(5.32)

which gives the asymptotic, providing an increase in the velocity when passing through the singular point. Eqs. (5.25)-(5.29) were numerically integrated on a uniform grid (with 10^7 mesh points from 0 to 1 along y) for T_0 = 4170 °K (Gingerich et al. 1971, 348) and $C = 3.8\times10^8$ cm^{-2} s^{-1}

(Feldman et al. 1977, 353-354). The mass flux value at the Earth orbit was obtained by averaging direct solar wind mass flux measurements from the "IMP 6" and the "IMP 8" spacecraft for three and a half years. The parameter n_0 was chosen proceeding from two conditions. First, the integral curve for the solar wind velocity should pass through the singular point. The nearest vicinity of the singular point was determined from the condition $|V^2 - T|T^{-1} < 10^{-5}$. The temperature in the vicinity of the singular point was calculated from Eq. (5.31), while the velocity was calculated from Eq. (5.32). Second, at the Earth orbit, the solar wind velocity and temperature should be close to their average values ($\langle V \rangle = 468$ km s^{-1}, $\langle T_e \rangle = 1.4\times10^5$ °K, and $\langle T_p \rangle = 1.2\times10^5$ °K) determined (Feldman et al. 1977, 353-354) from the same experimental data as those used here for the specification of the mass flux. The results of calculations at $n_0 = 0.85133\times10^{11}$ cm^{-3} are presented below. The temperature and the solar wind velocity at the Earth orbit are $T = 1.601\times10^5$ °K, $V = 511.856$ km s^{-1}, respectively. Fig. 5-5 shows the velocity and the temperature of the solar wind as functions of the heliocentric distance.

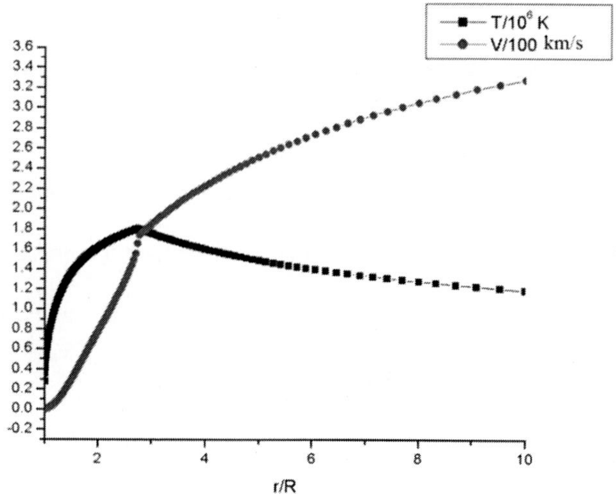

Fig. 5-5 The velocity and the temperature of the relic solar wind as functions of the heliocentric distance (adapted from Pisanko 2005, 428)

The velocity grows monotonically with the distance from the Sun. The temperature rapidly increases (by two orders of magnitude) from the photosphere to the base of the solar corona; it reaches its maximum (1.8×10^6 °K) at the singular point (r/R = 2.7491) and further decreases with the heliocentric distance. It means that the solar corona is heated by a sphere whose center coincides with the solar center and whose radius (r = 2.7491 R) is equal to the heliocentric distance to the singular point. The physical processes on this sphere, whose temperature is defined by Eq. (5.31) from the condition of the boundedness of the solar wind acceleration at this sphere, are beyond the model's scope. On the sphere, where $V^2 = T$ (i.e. the kinetic energy of the solar wind is equal to its thermal energy), favourable conditions can be expected for the turbulence development, hence, for the flow heating. If the mass flux of the solar wind is specified a priori, this automatically specifies the flux of the kinetic energy of its directed motion; the latter is the energy source for the coronal heating in the model. The process of the cascade energy transfer from large-scale to small-scale motions (down to the smallest scales) seems to occur in the solar wind. The viscosity in this process ultimately transforms the kinetic energy to the thermal one. In the model of the transonic flow of a non-viscid fluid, the only point where such an energy transformation via viscosity is "permissible" is the singular point. The viscosity is zero everywhere except for this point, where the flow is described by the asymptotic (which provides the boundedness of the flow acceleration) rather than by the model equations. Such a strong idealization results in a concentrated heat source at the singular point; this is an inevitable consequence of the neglect of the viscosity. The process of the viscous dissipation of energy is distributed over space in the actual solar wind but is concentrated at the singular point in the model. The expanding solar corona can be interpreted as a self-heating system: the heating of the solar corona can be related to the transonic regime of its expansion, which is maintained by the high temperature of the corona. A rapid decrease in temperature near the photosphere (Fig. 5-5) results from the terms describing energy losses due to the radiation: most of the coronal energy is lost via the thermal conduction down to the chromosphere, where the plasma is denser and, therefore, radiates more efficiently. The introduction of the recombination radiation means that neutral hydrogen atoms are formed. Since the model equations are written for the fully ionized hydrogen plasma, it is assumed that the neutral hydrogen atoms formed during the recombination are immediately re-ionized by collisions. At the hydrogen ionization potential of 13.54 eV, the hydrogen is ionized nearly completely even at temperatures of $\sim 10^5$ °K at low altitudes above the photosphere. It means that almost all neutral hydrogen is concentrated in the thin (compared to the solar radius) layer near the photosphere. The physics of this layer is complicated and is not the subject of the discussion here. This layer serves as the reservoir of protons and electrons, presenting there at comparatively low number densities. Since the gravitational attraction of electrons to the Sun is the weakest, they easily leave this thin layer to maintain the continuous plasma outflow to the interplanetary space, which is assumed in the model. The electric polarization field makes it possible to extract protons from the layer. This field does not affect the neutral hydrogen atoms. The temperature calculated at the Earth orbit ($T = 1.601\times10^5$ °K) closely matched the average of direct measurements ($\langle T_e \rangle = 1.4\times10^5$ °K, $\langle T_p \rangle = 1.2\times10^5$ °K) obtained (Feldman et al. 1977, 353-354) with the temperature measurement instrumental error of ±15% and the standard deviation of 0.4×10^5 °K for the electron temperature and 0.9×10^5 °K for the proton temperature. The solar wind velocity calculated at the Earth orbit (V = 511.856 km s^{-1}) proved to exceed the average result of the direct measurements ($\langle V\rangle$ = 468 km s^{-1} (Feldman et al. 1977, 353-354)) obtained with the velocity measurement instrumental error of ±2% and the standard deviation of 116 km s^{-1}. At the same time, within the instrumental error, this value is close to the average solar wind velocity of 504 km s^{-1} measured by the "Mariner 2" spacecraft during the solar minimum (Neugebauer and Snyder 1966, 4469-4484). Let us note that when Parker's theoretical results were first compared to observations on spacecraft, a so-called density problem was encountered, which is still urgent. The point is that the theory overestimates the interplanetary plasma density when the density at the coronal base is taken according to observations of Thompson's scattering. In the present model, the solar wind number density at the Earth orbit is 3.8×10^8 cm^{-2} s^{-1}/ 511.856×10^5 cm s^{-1} = 7.42 cm^{-3}. This value corresponds to the average $\langle n\rangle$ = 8.7 cm^{-3} obtained from direct measurements (Feldman et al. 1977, 353-354) with the density measurement instrumental error of ±30% and the standard deviation of 6.6 cm^{-3}. So, the constructed numerical solution links the electron number density in the minimum temperature region

(0.85133×10^{11} cm^{-3}) and that at the Earth orbit (7.42 cm^{-3}) and partly obviates the number density problem.

Flight to the Sun

The scientific community has been dreaming about the flight to the Sun for many years. Such a flight looks similar to pioneering flights to planets in the middle of the twentieth century when almost nothing was known about planets. Several actual scientific problems could be elucidated by observations on a specialized spacecraft.

First, one could descry tiny structured solar plasma elements with spatial scales of about 100 km and map the solar surface with the same image resolution. In spacecraft's polar trajectory, one could descry solar polar caps, almost invisible from the ecliptic. If a magnetograph were included in the payload, magnetic fields at solar polar caps would be measured.

Second, physical processes accelerating electrons up to a hundred MeV and protons up to a few GeV during a few seconds in solar flares remain enigmatic. More than 90% of accelerated particles are trapped into closed magnetic configurations situated in the solar atmosphere at heliocentric distances lower than three solar radii and never walk out to the interplanetary space. The presence of energetic particles inside magnetic traps is detected indirectly by observations of gamma, X-ray, microwave, and radio emissions and by detecting neutrons born in the solar atmosphere due to nuclear reactions initiated by energetic particles. The neutron lifetime is ~ 14 min because of its beta decay. Therefore, at the Earth orbit, one can register solar neutrons with energies higher than 100 MeV – the higher the neutron energy is, the faster it moves. On the spacecraft in the vicinity of the Sun, one could register solar neutrons with energies 1–10 MeV.

Third, until now, the nature of solar corona heating remains unclear. The small scale magnetic activity (like magnetic reconnections in the transition region and micro-flares in the chromospheric network) could generate ion cyclotron radio waves on frequencies up to a few kHz. Such waves can transmit their energy resonantly to coronal protons and heavier ions and heat it to very high temperatures. Observations of Doppler's broadenings of coronal emission lines with the help of UVCS and SUMER devices on the "SOHO" spacecraft testify to such a possibility. Because so high-frequency ion cyclotron waves cannot leave the solar corona to propagate in the interplanetary space, the waves can be registered only on the spacecraft flying in the solar corona, where the resonant wave-particle interaction could occur.

Fourth, direct measurements of the solar wind on the spacecraft flying in the solar corona along with remote sensing measurements of the solar wind onboard exo-atmospheric observatories ensure calibration tests of distant diagnostic means. If a Lyot coronograph is included in the payload, a series of K-corona photos making on the spacecraft in different projections along the trajectory will allow the tomographic reconstruction of the electron concentration three-dimensional structure of the corona.

Fifth, direct measurements of the dust on the spacecraft flying in the solar corona could elucidate the possible source of the coronal dust, whether it is the interplanetary dust approaching the Sun during its ten thousand-year motion along spiral solar centered orbits or cold material ejections from the solar interior. Because the maximum dust concentration region is situated at the heliocentric distance of about four solar radii, this could be investigated only on the spacecraft penetrated to such a distance into the solar corona.

Sixth, precise trajectory measurements of the near solar spacecraft could reveal weak trajectory deviations due to general relativity effects. It implies the compensation of the influence of the inhomogeneous solar heating of the spacecraft and its drag deceleration in the solar atmosphere and the other drifts responsible for non-inertial effects. The "Triad" spacecraft as an example of the drift-less satellite was already demonstrated in 1972. The Earth-directed special antenna should be mounted on the spacecraft to operate in X- and S- ranges to ensure the continuous Doppler control of the near solar spacecraft trajectory.

Engineering such a flight in the hostile environment of the solar corona creates problems. The minimum perihelion of the trajectory is limited by the necessity of the effective thermal shield of the spacecraft. For example, at the heliocentric distance of four solar radii, the solar heat flux is 400 watts per square centimetre, approximately three thousand times stronger than the solar constant. The material and the design of the thermal shield screen should minimize the gas evaporation from the screen to eliminate errors in onboard measurements of the solar wind plasma. At the same time, the solar corona creates problems in radio communications with the spacecraft. When the Sun is situated close to the directional pattern of the ground-based receiving station, the background noise is strongly enhanced, leading to the signal-to-noise merit significant decrease. To maximize the signal-to-noise merit, one should choose the trajectory so that the spacecraft would be seen from the Earth at the maximum possible angular distance up the angle of 90° at the perihelion. The coronal plasma creates the spacecraft signal scintillations that degrade the communication channel between the ground receiving station and the spacecraft.

Scintillation effects are minimized by choice of the carrier frequency as high as possible (X-range).

On August 12, 2018, the "Parker solar probe plus" spacecraft was launched into the circumsolar orbit with the perihelion at 9.5 solar radii, which is planned to be achieved by manoeuvring in Venus's gravitation field (Fox et al. 2016, 7-48). The mission duration is 6.9 years. The orbit lies almost in the ecliptic; its inclination is 3.4°. After initial maneuvering near Venus, the perihelion will be at 35 solar radii, while after the seventh one – at 9.5 solar radii. The mission will be declared successful after three successful orbits at 9.5 solar radii perihelion. The period of each such orbit is 88 days. There is the onboard thruster for manoeuvring, designed for "Messenger" and "Stereo" spacecraft. The tank for the hydrazine is taken from the "DCSC III" spacecraft. The scientific payload contains ion and electron analyzers, an energetic particle detector, a magnetometer, a plasma wave detector, a spectrometer of neutrons and gamma rays, a detector of dust and a hemispheric visualizer. The latter serves for binding direct plasma measurements to structural elements of K-corona. The ultimate scientific goal of the mission is to pioneer direct measurements of near solar environment characteristics in the ecliptic up the 9.5 solar radii heliocentric distances.

Is it possible to forecast solar wind gusts?

The question has no definite answer because of objective and subjective circumstances. The answer "yes" would be too optimistic in the spirit of the euphoria of this sort of project, while the answer "no" would look too pessimistic and even depressive.

Though the present-day Sun is much quieter than in the deep past when it was the rapidly rotating strongly magnetized super-active star with the intensive plasma wind, it can still create sporadic intense solar wind gusts called coronal mass ejections. In our intensive exploitation of the near-Earth space environment, solar wind gusts create more problems than that elucidated in mass media. The gusts can influence on operations of cellular phones, global positioning systems, meteorological satellites, electric energy distribution systems, wire phones, trunk pipelines etc.

Because nobody can reliably forecast when a coronal mass ejection will occur, one should be able to trace the ejection motion optically and (or) bear it by radio waves from its place of birth near the Sun to the Earth orbit. It needs precise measurements of ejection's velocity and its motion direction in every time moment.

Ground-based observations with the help of the Lyot coronograph cannot help in such monitoring because this device estimates only the electron concentration integral along a line-of-sight from the measured polarization brightness – the corona is optically thin. The scattered on coronal electrons solar light is polarized; therefore, one should consider the angles between the Sun and points of the emission in the corona and the observation at the Earth orbit. In the spherically symmetric corona, almost a half of the scattered light flux provides the sector situated at the angular distance of ±20° from the picture plane (the plane perpendicular to the line-of-sight), i.e. the Lyot coronograph is sensitive only to the part of the coronal mass ejection situated in the vicinity of the picture plane. Thus from ground-based coronograph observations, one can estimate only the projection of ejection's velocity on the picture plane. Such an estimate can differ from the actual ejection velocity significantly. The forecast of the ejection arrival from the Sun to the Earth based on terrestrial coronograph observations can lead to errors of a day or even more.

Stereoscopic cinematography is needed to monitor ejection's speed and its motion direction along its trajectory from the Sun. Two spacecraft situated at the Earth orbit observe the ejection under different angles ("The Sun and Heliosphere in Three Dimensions" – Report of the NASA Science Definition Team for the STEREO mission). One such spacecraft moves before the Earth and the other – behind the Earth. The position and the velocity of an ejection in every time moment can be reconstructed with the help of the classic geodesy technique along with the image recognition one when, first, the angular distance between spacecraft is known as well as angles of an ejection observation from every spacecraft and, second, an ejection is recognized. The latter is nontrivial because the image often is diffuse and has an irregular shape. Another way is the emitter location by onboard radio telescopes: the shock wave creating at the edge of an ejection generates radio emission in the kHz range and serves as an emitter. With the help of radio-telescopes mounted on two well-separated spacecraft, one could bear the moving emitter and reconstruct its trajectory by triangulation. Both optical and radio stereo monitoring improve the forecast accuracy of an ejection arrival up to a couple of hours. Because the onboard Lyot coronograph effectively observes K-corona at the angular distance of $\pm 20^{\circ}$ from the picture plane, both spacecraft should be separated at the angle of 90° at least. The triangulation is also most accurate at the same angular separation. For the stereo cinematography, an object should be photographed simultaneously with synchronism no worse than ± 1 s.

Two identical spacecraft with the common title “Stereo” were launched on October 26, 2006. Maneuvering in lunar gravitation was applied to displace one spacecraft at the Earth orbit to move ahead of the Earth and the other – behind the Earth. The onboard thrusters successfully operated on the “NOVA” spacecraft before. The scientific payload contains far ultraviolet and X-ray telescopes, the Lyot coronograph, the receiver of radio waves (similar to that mounted on the "Cassini" spacecraft) connected to the three-axis antenna system, an optical detector of variations of the solar light scattered on coronal electrons (with the angular resolution ~ 1° which is one hundred times better than in the device on the "Helios" spacecraft), the electrostatic analyzer of solar wind ions and electrons, the magnetometer, the detector of solar energetic electrons and ions. Each spacecraft is stabilized so that the Sun is inside the coronograph field of view in the range of ±30 arc seconds, and the image tremor should be no more than ±5 arc seconds. The rocking angle should be known with an accuracy no worse than 0.1° and is used to control the transmitting antenna direction to the Earth with the accuracy of ±0.1°. The information about the Sun swinging comes from star trackers mounted in directions perpendicular to the solar-terrestrial line because there are a lot of stars in the ecliptic, and these ensure the reliable recognition of the spacecraft swinging. The onboard computer calculates the direction to the Earth from the data on the spacecraft swinging and known elements of the spacecraft orbit. It determines the antenna tilt to maintain the reliable communication link with the Earth. The information is transmitted to the 34-meter in diameter antennas of the Deep Space Network once a day. The "Stereo" spacecraft information must be accessible in quasi-real-time to involve the stereo cinematography in the operative forecasting. For this purpose, the beacon transmission regime is foreseen. In case when the onboard computer automatically identified a coronal mass ejection in the coronograph image and the corresponding shock in radio receiver signals or the large flux of energetic particles, it preliminary estimates the speed and the direction of ejection's motion by analyzing the first few images and transmits the alarm signal to the Earth. The fastest ejections pass the distance of 5 solar radii during an hour; therefore, the picture frequency should register the ejection before it leaves the coronograph field of view. Moreover, the robotic recognition algorithm should maintain a relatively low level of false alarms. Furthermore, the more significant is the heliocentric distance, the more substantial F-corona background noise occurs.

Recurrent solar wind high-speed streams are other examples of solar wind gusts. Direct solar wind measurements on a spacecraft situated east to the solar-terrestrial line (i.e. moving behind the Earth) can be used in forecasting a high-speed stream arrival to the Earth with an advanced time of up to two days. The technique was tested on simultaneous data of "Pioneer 6-7-10-11" and "Vela 3-5" spacecraft. The correlation coefficient between the two datasets was 0.7 for the azimuthal distance between "Pioneer" and "Vela" spacecraft less than two days but fell to 0.28 for the four-day azimuthal distance. The forecast on two days was found the best possible one (Веселовский, 1984, 52-53).

Numerical calculations of the solar wind spatial structure and its inhomogeneities based on three-dimensional magneto-hydrodynamic models are considered possible means of forecasting high-speed stream arrivals to the Earth: the European initiative called EUHFORIA (EUropean Heliospheric FORecasting Information Asset) (Pomoell and Poedts 2018, https://doi.org/10.1051/swsc/2018020) was proposed. However, an attempt to exploit solutions of three-dimensional magneto-hydrodynamic equations for diagnostics (and, probably, forecasting) of high-speed streams in the solar wind looks thoughtless without reliable information about boundary conditions at the solar surface. The obligatory boundary conditions include the map of the photospheric magnetic field radial component and the three-dimensional geometry of magnetic confinement regions in the corona. The more reliable observational information one involves in the solution construction process, the less crude the diagnosis (or the forecast) is expected. Indeed, any models are based on reasonable assumptions about the solar wind properties, but models incorporating the maximum available reliable observational data look more credible. The three-dimensional geometry of magnetic confinement regions (manifested as helmet streamers, fans, and other structures of K-corona) could, in principle, be reconstructed from corona stereo images obtained with "Stereo" spacecraft. It was tested during the first year of the "Stereo" mission. Today only synoptic maps of the photospheric magnetic field line-of-sight component are available to use in global magneto-hydrodynamic models of the solar wind. To create a synoptic map, one should gather observational data approximately during a month. Therefore, the fidelity of the reproduction of high-speed streams with the help of global magneto-hydrodynamic models gives rise to reasonable doubts – the essential part of the initial information becomes approximately three weeks stale at the moment of the calculation of the solar wind spatial structure. Coronal observations on the "SOHO" spacecraft showed that the corona changed during hours and days than during weeks, even in quiet periods near the solar minimum. The relative success of the solar wind spatial structure global modelling in August-September 1996 ("Whole Sun Month") in the deep solar minimum was stipulated by fortunate circumstances. The simple ring of helmet streamers and fans around the solar equator with the only heliolongitude cut remained stable during almost a month (a total solar rotation) (Gibson et al. 1999, 871-879).

Elements of chance are natural for any forecasting. These elements cannot be eliminated but ought to be minimized based on reliable preliminary information. We suppose that deterministic forecasts of solar wind gusts look hardy possible due to a few reasons typical to weather forecasts. First, we poorly know all equations describing the system with sufficient accuracy. Second, we have no sufficient information on initial and boundary conditions that are obligatory in the deterministic description. Third, we neglect the "memory" of the system and therefore underestimate nonlinearities in the past and future. Forth, a horizon of forecasting and its accuracy remain unclear: both can be tested and explicitly formulated only post factum. Anyway, the diagnosis and forecasting need a reliable and well-thought-out communication system to alarm, modern technologies of monitoring, and common sense in interpreting the coming information about solar wind gusts.

In the first quarter of the twentieth century, Czech comic writer Yaroslav Hašek wrote in chapter XIII of his novel entitled "Adventures of brave soldier Shweik during the First World War.":

> "Some scientists explain the war by the occurrence of spots on the Sun. When such a spot appears, something terrible always occurs. The destruction of Carthage…", a soldier, the teacher by occupation, noted.
>
> "Save your education for yourself. What does this absurd spot mean? If there were even twenty such spots, these would be insufficient to get a greatcoat made", the corporal interrupted him.

"Once a spot appeared on the Sun. And I was beaten in the pub this day", Shweik broke into the conversation. "Since then, when I prepare to go anywhere, I see in newspapers whether there is any spot on the Sun. If the spot appears, I do not go and wait. When the volcano destroyed French island Martinique in the Atlantic Ocean, a professor wrote in "The National policy" Czech newspaper that he had alerted about the large spot on the Sun. Unfortunately, "The National policy" was not delivered to the island in time, and they got into a mess."

Summary of Chapter 5

So, let us summarize what we do not know about the solar wind but want to know.

Up to now, we do not know when and how the solar wind arose in the deep past and only guess that it could take place when the Sun came to the main sequence of its evolution and the nuclear fusion fired inside the Sun. We do not know when and how the solar wind will disappear in the future. We only guess that it would occur in case of the disappearance of the solar activity, which eventually maintains the hot solar corona, or when the Sun enters into the dense area of the local interstellar medium and/or into the powerful galactic magnetic field. Both scenarios are possible, but the conceivable time scale is the cosmogonic one.

We do not know how the solar wind originates close to the Sun and how it dissolves in the local interstellar medium. The solar wind velocity at the Earth orbit can variate in the broad interval, but one never observes values less than 200 km s^{-1} and more than 3000 km s^{-1}. Nobody knows why these limits exist and why the average solar wind density is several particles per cubic centimetre and not several thousands of particles per cubic centimetre or, conversely, a few particles per cubic metre. The evolution of these limits remains enigmatic – we observe the solar wind phenomenon for only seventy years.

We cannot satisfactorily explain why the helium abundance in the solar wind at the Earth orbit does not exceed 25–30% relative to protons and why rich abundances of heavy ions have never been observed despite many separation mechanisms.

The role of solar radiative emission remains still unclear. The pressure of light could both push the plasma from the Sun and resist this elevation process. One usually considers the plasma (where the light propagates) either optically thin or optically thick. The first choice implies that the solar light propagation and the plasma motion are uncoupled. The second one needs to take into account the complex energy redistribution between the plasma and the light. There is the commonly accepted point of view that solar light does not influence the solar wind. The solar wind mass flux is often established as an additional external condition in solar wind models. The fact that the flux is of the order of the total rest mass of the solar material taken away with the solar light is interpreted as a random coincidence. The current rate of solar mass losses due to the solar wind is estimated as 2×10^{18} g s^{-1}, i.e. of the order of 10^{-14} of the solar mass per year.

Until now, the nature of solar corona heating remains unclear. Small scale magnetic activities (like magnetic reconnections in the transition region and microflares in the chromosphere network) could generate ion cyclotron radio waves on frequencies up to a few kHz. Such waves can transmit their energy resonantly to coronal protons and heavier ions and heat it to very high temperatures. Because so high-frequency ion cyclotron waves cannot leave the solar corona to propagate in the interplanetary space, the waves can be registered only on the spacecraft flying in the solar corona.

BIBLIOGRAPHY

References to Chapter 1

Виткевич, В.В. 1951. "Новый метод исследования солнечной короны". *Доклады АН СССР*. т.77, №4. 585-588 (in Russian)

Виткевич, В.В. 1955. "Результаты наблюдений рассеяния радиоволн на электронных неоднородностях солнечной короны". *Доклады АН СССР*. т.101, №3. 429-432 (in Russian)

Виткевич, В.В., Антонов, Т.Д., Власов, В.И. 1966. "Наблюдения флуктуаций интенсивности радиоизлучения квазизвездного источника 3C48 на неоднородностях межпланетной плазмы". *Доклады АН СССР*. т.168, №1. 55-59 (in Russian)

Всехсвятский, С.К., Никольский, Г.М., Пономарёв, Е.А., Чередниченко, В.И. 1955. "К вопросу о корпускулярном излучении Солнца". *Астрономический журнал*. т.32. 165-176. (in Russian)

Грингауз, К.И., Безруких, В.В., Озеров, В.Д., Рыбчинский, Р.Е. 1960. "Изучение межпланетного ионизированного газа, энергичных электронов и корпускулярного излучения Солнца при помощи трёхэлектродных ловушек заряженных частиц на второй советской космической ракете". *Доклады АН СССР*. т.131. 1301-1304. (in Russian)

Ahnert, P. 1943. "Der komet 1942 g (Whipple-Fedtke)". *Zeitschrift fur Astrophysik*. v.22. 288-309

Bartels, J. 1932. "Terrestrial-magnetic activity and its relations to solar phenomena". *Terr. Mag. Atmos. Electr.* v.37. 1-52

Biermann, L. 1951. "Kometenschweife und solare korpuskularstrahlung". *Zeitschrift fur Astrophysik*. v.29. 274-286

Biermann, L. 1953. "Physical processes in comet tails and their relation to solar activity". *Extrait des Mem.Soc.Roy.Sci.Liege Collection in -4°*. v.13. 291-302

Biermann, L. 1957. "Solar corpuscular radiation and the interplanetary gas". *Observatory*. v.77. 109-110

Bonetti, A., Bridge, H.S., Lazarus, A.J., Lyon, E.F., Rossi, R., Scherb, F. 1963. "Explorer 10 plasma measurements". *J. Geophys. Res.* v.68. 4017-4063

Bredichin, Th. 1861. "Quelques mots sur les queues des cometes". *Die Astronomische Nachrichten*. Bd. 1291

Bredichin, Th. 1861. "Sur la direction des queues des cometes de 1844, III et de 1853, III.". *Die Astronomische Nachrichten*. Bd.1305, Bd.1306

Bredichin, Th. 1878. "Sur la force qui produit les queues des cometes". *Die Astronomische Nachrichten*. Bd. 93.

Chamberlain, J.W. 1961. "Interplanetary gas, III. A hydrodynamic model of the corona". *Astrophys. J.* v.133. 675-687

Chapman, S. 1919. "An outline of a theory of magnetic storms". *Proc. Roy. Soc. Lond.* A95. 61-83

Chapman, S., Ferraro, V.C.A. 1931. "A new theory of magnetic storms, Part I, The initial phase". *Terr. Magn. Atmos. Elect.* v.36. 77-97

Chapman, S., Ferraro, V.C.A. 1933. "A new theory of magnetic storms". *Terr. Magn. Atmos. Elect.* v.38. №79. 171-196

Chapman, S., Ferraro, V.C.A. 1940. "The theory of the initial phase of a geomagnetic storm". *Terr. Magn. Atmos. Elect.* v.45. 245-268

Chapman, S. 1957. "Notes on the solar corona and the terrestrial ionosphere". *Smithsonian Contrib. Astrophys.* v.2. 1-11

Dennison, P.A., Hewish, A. 1967. "The solar wind outside the plane of ecliptic". *Nature.* v.213. 343-344

Edlen, B. 1943. "Die duetung der emissionslinien im spektrum der sonnenkorona". *Zeitschrift fur Astrophysik.* v.22. 30-63

FitzGerald, G.F. 1900. "Sunspots, magnetic storms, comet tails, atmospheric electricity, and aurorae". *The Electrician.* v.46. 287-288

Forbush, S.E. 1946. "Three unusual cosmic-ray increases possibly due to charged particles from the Sun". *Phys. Rev.* v.70. №9-10. 771-772

Gringauz, K.I. 1961. "Some results of experiments in interplanetary space by means of charged particle traps on Soviet space probes". *Space Res.* v.2. 539-553

Hewish, A. 1955. "The irregular structure of the outer regions of the solar corona". *Proc. Roy. Soc.* v.228. №1173. 238-251

Hewish, A., Scott P., Wills D. 1964. "Interplanetary scintillation of small diameter radio sources". *Nature.* v.203. №4950. 1214-1217

von Hoffmeister, C. 1943. "Physikalische untersuchungen an kometen. I. Die beziehungen des primaren schweifstrahls zum radiusvektor". *Zeitschrift fur Astrophysik.* v.22. 265-287

Lodge, O. 1900. "Sunspots, magnetic storms, comet tails, atmospheric electricity and aurorae". *The Electrician.* №46. 249-250

Neugebauer, M., Snyder, C.W. 1966. "Mariner 2 observations of the solar wind, 1. Average properties". *J. Geophys. Res.* v.71. 4469-4484

Parker, E.N. 1958. "Dynamics of the interplanetary gas and magnetic fields". *Astrophys. J.* v.128. 644-675

Scherb, F. 1964. "Velocity distributions of the interplanetary plasma detected by Explorer 10". *Space Res.* v.4. 797-818

Snyder, C.W., Neugebauer, M. 1964. “Interplanetary solar wind measurements by Mariner-2”. *Space Res*. v.4. 89-113

References to Chapter 2

Арманд, Н.А., Ефимов, А.И., Яковлев, О.И., Якубов, В.П., Вышлов, А.С. 1988. “Флуктуации частоты дециметровых и сантиметровых радиоволн при связи с аппаратами Венера-15,16 через околосолнечную плазму”. *Радиотехника и электроника*. т.33. №8. 1574-1583 (in Russian)

Ефимов, А.И., Яковлев, О.И., Разманов, В.М., Штрыков, В.К. 1977. “Турбулентность околосолнечной плазмы и скорость солнечного ветра по данным радиопросвечивания с помощью аппарата Венера-10”. *Письма в Астрономический журнал*. т.3. №7. 322-325 (in Russian)

Ефимов, А.И., Яковлев, О.И., Штрыков, В. К. Рогальский, В.И., Тихонов, В.Ф. 1981. “Разнесенные наблюдения флуктуаций частоты и фазы радиоволн, рассеянных околосолнечной плазмой”. *Радиотехника и электроника*. т.26. №2. 311-318 (in Russian)

Ефимов, А.И., Яковлев, О.И., Рубцов, С.Н., Кучерявенков, А.И., Молотов, Е.И., Рогальский, В.И., Якубов, В.П., Кафтонов, А.С., Хажеев, Р.С. 1987. “Временные спектры флуктуаций амплитуды сантиметровых и дециметровых радиоволн аппаратов «Венера-15 и 16» при просвечивании околосолнечной плазмы”. *Радиотехника и электроника*. т.32. №10. 2025-2032 (in Russian)

Ефимов, А.И., Чашей, Н.В., Шишов, В.Н., Яковлев, О.И. 1990. “Ускорение солнечного ветра по данным радиопросвечивания”. *Космические исследования*. т.28. №4. 581-587. (in Russian)

Ефимов, А.И., Рудаш, В.К., Чашей, И.В., Яковлев, О.И. 1993. “О детектировании МГД - волн в околосолнечной плазме методом радиопросвечивания”. *Астрономический журнал*. т.70. №5. 1075-1085 (in Russian)

Ефимов, А.И., Чашей, И. В., Шишов, В.И., Берд, М.К. 1993. “О флуктуациях фарадеевского вращения при радиопросвечивании околосолнечной плазмы”. *Письма в Астрономический журнал*. т.19. №2. 143-152 (in Russian)

Ефимов, А.И., Андреев, В.Е., Самознаев, Л.Н., Чашей, А.И., Берд, М.К. 1999. “Влияние корональных магнитных полей на формирование солнечного ветра по данным поляризационного радиопросвечивания”. *Астрономический журнал*. т.76. №4. 312-320 (in Russian)

Ефимов, А.И., Чашей, Н.В., Самознаев, Л.Н., Андреев, В.Е., Берд, М.К., Еденхофер, П., Плеттемейер, Д., Вольмут, Р. 2002. “Внешний масштаб турбулентности солнечного ветра по данным радиозатменных экспериментов с использованием космического аппарата Galileo”. *Астрономический журнал*. т.79. №7. 640-650 (in Russian)

Ефимов, А.И., Арманд, Н.А., Луканина, Л.А., Самознаев, Л.Н., Чашей, И.В., Берд, М.К., Плеттемейер, Д. 2008. “Радиальная зависимость уровня амплитудных флуктуаций сигналов космических аппаратов при зондировании околосолнечной плазмы”. *Радиотехника и электроника*. т.53. №10. 1257-1268 (in Russian)

Ефимов, А.И., Луканина, Л.А., Самознаев, Л.Н., Чашей, И.В., Берд, М.К., Плеттемейер, Д. 2009. “Анализ частотных флуктуаций радиоволн в околосолнечной плазме по данным аппарата Galileo”. *Радиотехника и электроника*. т.54. №7. 773-784. (in Russian)

Ефимов, А.И., Луканина, Л.А., Самознаев, Л.Н., Чашей, И.В., Берд, М.К. 2010. “Интенсивность флуктуаций частоты радиосигналов аппаратов в околосолнечной плазме”. *Радиотехника и электроника*. т.55. №11. 1343-1349 (in Russian)

Ефимов, А.И., Луканина, Л.А., Самознаев, Л.Н., Чашей, И.В., Берд, М.К., Плеттемейер, Д. 2010. “О пространственном распределении характеристик турбулентности во внутреннем солнечном ветре”. *Астрономический журнал*. т.87. №5. 492-502 (in Russian)

Ефимов, А.И., Луканина, Л.А., Рогашкова, А.И., Самознаев, Л.Н., Чашей, И.В., Берд, М.К., Петцольд, М. 2011. “Квазипериодические возмущения в околосолнечной плазме при двухчастотном радиозондировании сигналами космических аппаратов”. *Радиотехника и электроника*. т.56. №12. 1433-1443 (in Russian)

Ефимов, А.И., Луканина, Л.А., Рудаш, В.К., Самознаев, Л.Н., Чашей, И.В., Берд, М.К., Петцольд, М. 2013. “Частотные флуктуации когерентных сигналов космических аппаратов при радиозондировании околосолнечной плазмы в 2004-2008 г.”. *Космические исследования*. т.51. №1. 17-27 (in Russian)

Ефимов, А.И., Луканина, Л.А., Рогашкова, А.И., Самознаев, Л.Н., Чашей, И.В., Берд, М.К., Петцольд, М. 2013. “Квазипериодические флуктуации фарадеевского вращения плоскости поляризации при радиозондировании околосолнечной плазмы”. *Радиотехника и электроника*. т.58. №9. 916-925 (in Russian)

Колосов, М.А., Яковлев, О.И., Ефимов, А. И., Разманов, В.М., Штрыков, В.К. 1978. “Исследование флуктуаций радиоволн и турбулентности околосолнечной плазмы в экспериментах радиопросвечивания с помощью аппарата Венера-10”. *Доклады АН СССР*. т.241. №3. 555-558 (in Russian)

Колосов, М.А., Яковлев, О.И., Ефимов, А.И., Молотов, Е.П., Рогальский, В.И., Разманов, В.М., Штрыков, В.К. 1978. “Исследование распространения дециметровых радиоволн в околосолнечной плазме при полете межпланетной станции Венера-10”. *Радиотехника и электроника*. т.23. №9. 1829-1839 (in Russian)

Лотова, Н.А. 1968. “Радиоастрономические исследования неоднородной структуры околосолнечной плазмы”. *Успехи физических наук*. т.95. №2. 292-307 (in Russian)

Разманов, В.М., Ефимов, А.И., Яковлев, О.И. 1979. “О формировании спектральной линии радиосигналов при распространении в околосолнечной плазме”. *Известия вузов. Радиофизика*. т.22. №9. 1051-1060 (in Russian)

Рубцов, С.Н., Яковлев, О.И., Ефимов, А.И. 1987. “Концентрация, неоднородность плазмы и кинетическая энергия солнечного ветра по данным радиопросвечивания с использованием аппаратов Венера-15 и 16”. *Космические исследования*. т.25. №4. 620-625. (in Russian)

Яковлев, О.И., Трусов, Б.П., Виноградов, В.А., Ефимов, А.И. Круглов, Ю.М., Матюгов, С.С., Разманов, В.М. 1974. “Исследование распространения дециметровых радиоволн в околосолнечной плазме с помощью аппарата Марс-2”. *Космические исследования*. т.12. №4. 600-605 (in Russian)

Яковлев, О.И., Молотов, Е.П., Ефимов, А.И., Круглов, Ю.М., Разманов, В.М., Тимофеева, Т.С., Штрыков, В.К. 1977. “Особенности распространения радиоволн через околосолнечную плазму по данным аппаратов Марс-2 и 7”. *Радиотехника и электроника*. т.22. №2. 260-267 (in Russian)

Яковлев, О.И., Ефимов, А.И., Разманов, В.М., Штрыков, В.К. 1980. “Неоднородная структура и скорость движения околосолнечной плазмы по данным станции Венера-10”. *Астрономический журнал*. т.57. №4. 790-798 (in Russian)

Яковлев, О.И., Ефимов, А.И., Рубцов, С.Н. 1987. “Динамика и турбулентность солнечного ветра в области его формирования по данным радиопросвечивания с применением аппаратов Венера-15 и 16”. *Космические исследования*. т.25. №2. 251-257 (in Russian)

Яковлев, О.И., Ефимов, А.И., Молотов, Е.П., Рубцов, С.Н., Якубов, В.П., Кучерявенков, А.И., Кафтонов, А.С. 1988. “Флуктуации амплитуды дециметровых и сантиметровых радиоволн, излучаемых спутниками Венера-15 и 16 при распространении через околосолнечную плазму”. *Известия вузов. Радиофизика*. т.31. №1. 3-9 (in Russian)

Яковлев, О.И., Ефимов, А.И., Рубцов, С.Н. 1988. “Солнечный ветер по данным радиопросвечивания с помощью аппаратов Венера-15 и 16”. *Астрономический журнал*. т.65. №6. 1290-1299 (in Russian)

Яковлев, О.И., Ефимов, А.И., Якубов, В.П., Корсак, О.М., Кафтонов, А.С., Ерофеев, А.Л., Рубцов, С.Н. 1989. “Флуктуации частоты и фазы радиоволн в двух разнесенных пунктах при просвечивании околосолнечной плазмы и скорость солнечного ветра”. *Известия вузов. Радиофизика*. т.32. №5. 531-537 (in Russian)

Яковлев, О.И. 2014. “Определение скорости, концентрации, мощности солнечного ветра методом радиопросвечивания сигналами космических аппаратов”. *Известия вузов. Радиофизика*. т.57. №5. 347-361 (in Russian)

Яковлев, О.И. 2017. “Характеристики турбулентности и волн околосолнечной плазмы по данным радиопросвечивания сигналами космических аппаратов”. *Известия вузов. Радиофизика*. т.60. №4. 289-303 (in Russian)

Якубов, В.П., Яковлев, О.И., Ефимов, А.И., Ерофеев, А.Л., Корсак, О.М., Ерофеева, М.А. 1989. “Волны плотности в околосолнечной плазме по данным радиопросвечивания”. *Космические исследования*. т.27. №5. 772-775 (in Russian)

Якубов, В.П., Яковлев, О.И., Ефимов, А.И., Ерофеев, А.П. 1991. “Скорость солнечного ветра по частотным данным двукратного радиопросвечивания околосолнечной плазмы”. *Известия вузов. Радиофизика*. т.34. №6. 615-623 (in Russian)

Anderson, J.D., Krisher, T.P., Borutzki, S.E., Connally, J., Eshe, P., Moyd, K., Roth, D., Sweenam, D.C., Taylor, A.H., Tyler, G.L., Gresh, D.L., Rosen, A. 1987. “Radio range measurements of coronal electron densities during the solar conjunction of Voyager-2”. *Astrophys. J.* v.323(2). L141-L143

Armand, N.A., Efimov, A.I., Yakovlev, O.I. 1987. “A model of the solar wind turbulence from radio occultation experiments”. *Astron. Astrophys*. v.183. 135-141

Armstrong, J.W., Woo, R. 1981. "Solar wind motion within 30R, spacecraft radioscintillation observations". *Astron. Astrophys.* v.103. 415-421

Bird, M.K. 1982. "Coronal investigations with occulted spacecraft signals". *Space Sci. Rev.* v.33. 99-126

Bird, M.K., Volland, H., Patzold, M., Edenhofer, P., Asmar, S.W., Brenkle, J.P. 1994. "The coronal electron density distribution determined from dual-frequency ranging measurements during the 1991 solar conjunction of the Ulysses spacecraft". *Astrophys. J.* v.426(1). 373-381

Bird, M.K., Patzold, M., Edenhofer, P., Asmar, S.W., Mcelrath, T.P. 1996. "Coronal radio sounding with Ulysses: solar wind electron density during the 1995 conjunction". *Astron. Astrophys.* v.316. 441-448

Chashei, I.V., Efimov, A.I., Samoznaev, L.N., Plettemeier, D., Bird, M.K. 2005. "Properties of solar wind turbulence near the sun as deduced from coronal radio sounding". *Adv. Space Res.* v.36(8). 1454-1460

Chashei, I.V., Efimov, A.I., Bird, M.K. 2007. "Solar wind turbulence from radio occultation data". *Astron. Astrophys. Trans.* v.26(6). 611-620

Dollfus, A., Mouradian, Z. 1981. "Electrons in the solar corona". *Solar Phys.* v.70. №1. 3-10

Efimov, A.I., Yakovlev, O.I., Razmanov, V.M., Shtrykov, V.K. 1977. "Turbulence of plasma near the sun and the solar wind velocity derived from the Venera 10 radiooccultation experiment". *Astronomy Letters.* v.3(7). 172-180

Efimov, A.I., Chashej, I.V., Shishov, V.I., Yakovlev, O.I. 1994. "Detection of wave phenomena in the solar corona using radio occultation method". *Adv. Space Res.* v.14(4). 493-496

Efimov, A. I., Samoznaev, L.N., Andreev, V. E., Bird, M.K., Edenhofer, P., Plettemeier, D., Wohlmuth, R. 2002. "East-west scattering level asymmetry of the solar corona". *Adv. Space Res.* v.30(3). 453-458

Efimov, A.I., Chashei, I.V., Samoznaev, L.N., Andreev, V.E., Edenhofer, P., Plettemeier, D., Wohlmuth, R. 2002. "The outer scale of solar wind turbulence from Galileo coronal-sounding data". *Astronomy Reports*. v.46(7). 579-590

Efimov, A.I., Bird, M.K., Chashei, I.V. 2003. "Simultaneous observations of radio wave frequency and intensity fluctuations for estimates of solar wind speed". *Adv. Space Res.* v.32(4). 485-490

Efimov, A.I., Chashei, I.V., Bird, M.K., Plettemeier, D., Edenhofer, P., Wohlmuth, R. 2005. "Turbulence of the inner solar wind at solar maximum: coronal radio sounding with Galileo in 1999". *Adv. Space Res.* v.36. 1448-1453

Efimov, A.I., Bird, M.K., Rudash, V.K., Andreev, V.E., Chashei, I.B., Plettemeier, D., Edenhofer, P. 2005. "Solar wind velocity measurements near the sun using Ulysses radio amplitude correlation at two frequencies". *Adv. Space Res.* v.35. 2189-2194

Efimov, A.I., Samoznaev, L.N., Bird, M.K., Chashei, I.V., Plettemeier, D. 2008. "Solar wind turbulence during the solar cycle deduced from Galileo coronal radio-sounding experiments". *Adv. Space Res.* v.42(2). 117-123

Efimov, A.I., Lukanina, L.A., Samoznaev, L.N., Rudash, V.K., Chashei, I.B., Bird, M.K., Patzold, M. 2012. "Quasi-periodic frequency fluctuations observed during coronal radio sounding experiments 1991-2009". *Adv. Space Res.* v.49. 500-508

Efimov, A.I., Lukanina, L.A., Samoznaev, L. N., Rudash, V.K., Chashei, I.V., Bird, M.K., Patzold, M. 2014. "Two-way frequency fluctuations observed during coronal radio sounding experiments". *Solar Phys.* v.289(5). 1715-1729

Efimov, A.I., Lukanina, L.A., Rogashkova, A.I., Samoznaev, L.N., Chashei, I.V., Bird, M.K., Patzold, M. 2015. "Coronal radio occultation experiments: analysis of Faradey rotation fluctuations". *Solar Phys.* v.290(9). 2397-2408

Efimov, A.I., Lukanina, L.A., Samoznaev, L.N., Chashei, I.V., Bird, M.K., Paetzold, M. 2017. "Frequency fluctuations in the solar corona investigated with radio sounding experiments on the spacecraft ROSETTA and MARS EXPRESS in 2010/2011". *Adv. Space Res*. v.59. 1652-1662

Erickson, W.C. 1964. "The radio-wave scattering properties of the solar corona". *Astrophys. J.* v.139. 1290-1311

Goldstein, R.M. 1969. "Superior conjunction of Pioneer-6". *Science*. v.166. 598-602

Hewish, A., Symonds, M. 1969. "Radio investigation of the solar plasma". *Planet Space Sci*. v.17. №3. 313-329

Hollweg, J.V. 1968. "Solar coronal effect on pulsar signals". *Nature*. v.220. №23. 771-779

Imamura, T., Tokumaru, M., Hiroaki, I., Isobe, H., Shiota, D., Ando, H., Miyamoto, M., Toda, T. 2014. "Outflow structure of the quiet sun corona probe by spacecraft radio scintillation in strong scattering". *Astrophys. J.* v.788. 117-135

Janardhan, P., Bird, M.K., Edenhofer, P., Wohlmuth, R., Plettemeier, D., Asmar, S.W., Patzold, M., Karl, J. 1999. "Coronal velocity measurements with Ulysses: multi-link correlation studies during two superior conjunctions". *Solar Phys*. v.184(1). 157-172

Jensen, E.A., Nolan, M., Bisi, M., Chashei, I., Vilas, F. 2013. "Messenger observations of magnetohydrodinamic waves in the solar corona from Faraday rotation". *Solar Phys*. v.285. 71-82

Jensen, E.A., Frazin, R., Heiles, C., Lamy, P., Llebaria, A., Anderson, J.D., Bisi, M.M., Fallows, R.A. 2016. "The comparison of total electron content between radio and Tompson scattering". *Solar Phys*. v.291. 465-485

James, J.C. 1970. "Some observed characteristics of solar radar echoes and their implications". *Solar Phys*. v.12. 143-162

Krisher, T.P., Anderson, J.D., Morabito, D.D., Asmar, S.W., Borutzki, S.E., Delitsky, M.L., Densmore, A.C., Eshe, P.M., Lewis, G.D., Maurer, M.J., Roth, D.C., Son, Y.H., Spilker, T.R., Sweetnam, D.N., Taylor, A.H., Tyler, G.L., Gresh, D.L., Rosen, P.A. 1991. "Radio range measurements of coronal electron densities at 13 and 3.6 centimeter wavelengths during 1988 solar conjunction". *Astrophys. J.* v.375(1). L57-L61

Kolosov, M.A., Yakovlev, O.I., Efimov, A.I., Rogalsky, V.I., Razmanov, V.M., Shtrykov, V.K. 1982. "Decimeter radio wave propagation in the turbulent plasma near the sun, using Venera10 spacecraft". *Radio Sci.* v.17(3). 664-674

Levy, G.S., Sato, T., Seidel, B.L., Steilzried C.T. 1999. "Pioneer-6: measurement of transient Faraday rotation phenomena observed during solar occultation". *Science.* v.166. №3905. 596-600

Miyamoto, M., Imamura, T., Tokumaru, M., Ando, H., Isobe, H., Asai, A., Shiota, D., Toda, T., Hausler, B., Patzold, M., Nabatov, A., Nakamura, M. 2014. "Radial distribution of compressive waves in the solar corona revealed by Akatsuki radio occultation observations". *Astrophys. J.* v.797. 51-58

Muhleman, D.O., Esposito, P.B., Anderson, J.D. 1977. "The electron density profile of the outer corona and the interplanetary medium from Mariner-6 and Mariner-7 time delay measurements". *Astrophys. J.* v.211(3). 943-957

Muhleman, D.O., Anderson, J.D. 1981. "Solar wind electron densities from Viking dual-frequency radio measurements". *Astrophys. J.* v.247(3). 1093-1101

Ness, F.N., Wilcox, J.M. 1965. "Sector structure of the quiet interplanetary magnetic field". *Science.* v.147. №3677. 1592-1594

Patzold, M., Bird, M.K., Edenhofer, P., Asmar, S.W., McElrath, T.P. 1995. "Dual-frequency radio sounding of the solar corona during the 1995 conjunction of the Ulysses spacecraft". *Geophys. Res. Lett.* v.22(23). 3313-3317

Patzold, M., Bird, M.K. 1998. “Polar plumes and fine-scale coronal structures of radio sounding data”. *Geophys. Res. Lett.* v.25. 1845-1849

Patzold, M., Bird, M.K., Volland, H., Levy, S.G., Seidel, B.L., Stelzried, C.T. 1987. “The mean coronal magnetic field determined from Helios Faraday rotation measurements”. *Solar Phys.* v.109. №1. 91-105

Paetzold, M., Hahn, M., Tellmann, S., Hausler, S., Bird, M.K., Tyler, G.L., Asmar, S.W., Tsurutani, B.T. 2012. “Coronal density structures and CMEs: superior solar conjunctions of Mars Express, Venus Express, and Rosetta: 2004, 2006, and 2008”. *Solar Phys.* v.279. 127-152

Pisanko, Yu.V. 1996. “The global solar magnetic field as a “controller” for the origin and acceleration model of solar wind streams”. *Adv. Space Res.* v.17(3). (3)61-(3)64

Stelzried, T.C., Levy, G.S., Sato, T., Rusch, W.V.T., Ohlson, J.E., Schatten, K.H., Wilcox, J.M. 1970. “The quasi-stationary coronal magnetic field and electron density as determined from a Faraday rotation experiment”. *Solar Phys.* v.14. №2. 440-456

Tyler, G.L., Vesecky, J.F., Plume, M.A., Howard, H.T., Barnes, A. 1981. “Radio wave scattering observations of the solar corona: first-order measurements of expansion velocity and turbulence spectrum using Viking and Mariner-10 spacecraft”. *Astrophys. J.* v.249. 318-332

Volland, H., Bird, M.K., Levy, G.S., Stelzried, C.T., Seidel, B.L. 1977. “Helios-1 Faraday rotation experiment: results and interpretations of the solar occultations in 1975”. *Journal of Geophysics - Zeitschrift fur Geophysik.* v.42. №6. 659-672

Weisberg, J.M., Rankin, J.M., Payne, R.R., Counselman III C.C. 1976. “Further changes in the distribution of density and radio scattering in the solar corona in 1973”. *Astrophys. J.* v.209. 252-258

Wohlmuth, R., Plettemeier, D., Edenhofer, P., Bird, M.K., Patzold, M., Asmar, S.W. 1997. “Measurement of the propagation speed of plasma inhomogeneities in the solar corona using an uplink/downlink cross – correlation method". *Radio Sci.* v.32(2). 617-628

Woo, R., Yang, F.C., Ishimaru, A. 1976. "Structure of density fluctuations near the sun deduced from Pioneer-6 spectral broadening measurements". *Astrophys. J.* v.210. 593-602

Woo, R., Yang, F.C., Yip, K., Kendall, W.B. 1976. "Measurements of large-scale density fluctuations in the solar wind using dual-frequency phase scintillations". *Astrophys. J.* v.210. 568-574

Woo, R. 1978. "Radial dependence of solar wind properties deduced from Helios-1,2 and Pioneer-10,11 radio scattering observations". *Astrophys. J.* v.219. 727-739

Woo, R., Armstrong, J.W., Bird, M.K., Patzold, M. 1995. "Variation of fractional electron density fluctuations using 40R observed by Ulysses ranging measurements". *Geophys. Res. Lett.* v.22(4). 329-332

Woo, R. 1996. "1985 Voyager-2 radio ranging measurements of coronal density". *Astrophys. J.* v.458. L87-L90

Yakovlev, O.I., Efimov, A.I., Razmanov, V.M., Shtrykov, V.K. 1980. "Radio wave propagation in the turbulent solar plasma using three satellites". *Acta astronautica*. v.7. 235-242

Yakovlev, O.I. 2002. *Space Radio Science*. London: Taylor - Francis.

Yakovlev, O.I., Pisanko, Y.V. 2018. "Radio sounding of the solar wind acceleration region with spacecraft signals". *Adv. Space Res.* v.61. №1. 552-566

Yakovlev, O.I., Yakovlev, Y.O. 2019. "Analysis of rapid dependencies of the electron concentration and plasma velocity in the region of accelerated solar wind". *Cosmic Research*. v.57 (4). 237-242

References to Chapter 3

Баранов, В.Б., Краснобаев, К.В. 1977. *Гидродинамическая теория межпланетной плазмы*. Москва: Наука (in Russian)

Веселовский, И.С. 1984. *Физика межпланетной плазмы. Итоги науки и техники. Серия «Исследование космического пространства»*, т.22. Москва: ВИНИТИ (in Russian)

Ермолаев, Ю.И. 2002. "Экспериментальное изучение крупномасштабной структуры солнечного ветра". Диссертация на соискание учёной степени доктора физико-математических наук, ИКИ РАН (in Russian)

Aellig, M., Lazarus, A., Steinberg, J. 2001. "The solar wind helium abundance: variation with wind speed and the solar cycle". *Geophys. Res. Lett*. v.28. №14. 2767-2770

Bale, S.D., Kasper, J.C., Howes, G.G., Quataert, E., Salem, C., Sundkvist, D. 2009. "Magnetic fluctuation power near proton temperature anisotropy instability thresholds in the solar wind". *Phys. Rev. Lett*. v.103. 211101-211102

Balogh, A., Smith, E.J., Tsurutani, B.T., Southwood, D.J., Forsyth, R.J., Hornbury. 1995. "The heliospheric magnetic field over the south polar region of the sun". *Science*. v.268. 1007-1010

Balogh, A., Forsyth, R.J. 1998. "The results of the Ulysses mission: a survey of the heliosphere in three dimensions". In *A Crossroads for European Solar and Heliospheric Physics*, edited by E.R. Priest, F. Moreno-Insertis, R.A. Harris, 45-55. ESA SP-417

Bame, S.J., Asbridge, I.R., Feldman, W.C., Montgomery, M.D., Kearney, P.D. 1975. "Solar wind heavy ion abundances". *Solar Phys*. v.43. 463-473

Beinroth, H.J., Neubauer, F.M. 1981. "Properties of whistler mode waves between 0.3 and 1.0 AU from "Helios" observations". *J. Geophys. Res*. v.A86. №9. 7755-7760

Belcher, J. and Davis, L., Jr. 1971. "Large-amplitude Alfven waves in the interplanetary medium, 2". *J. Geophys. Res*. v.76. №16. 3534-3563

Bieber, J.W., Chen, J., Matthaeus, W.H., Smith, C.W., Pomerantz, M.A. 1993. "Long-term variations of interplanetary magnetic field spectra with implications for cosmic ray modulation". *J. Geophys. Res*. v.A3. 3585-3603

Borrini, G., Gosling, J.T., Bame, S.J., Feldman, W.C., Wilcox, J.M. 1981. "Solar wind helium and hydrogen structure near the heliospheric current sheet: a signal of coronal streamers at 1 AU". *J. Geophys. Res*. v.86. №A6. 4565-4573

Borrini, G., Gosling, J.T., Bame, S.J., Feldman, W.C. 1982. "Helium abundance enhancements in the solar wind". *J. Geophys. Res*. v.87. №A9. 7370-7378

Bothmer, V., Schwenn, R. 1995. "The interplanetary and solar causes of major geomagnetic storms". *J. Geomagn. Geoelectr*. v.47. 1127-1132

Bruno R., Carbone V. 2005. "The solar wind as a turbulence laboratory". *Living Rev. Sol. Phys*. v.2. 4-185

Burlaga, L.F., Behannon, K., Klein, L. 1987. "Compound streams, magnetic clouds, and major geomagnetic storms". *J. Geophys. Res*. v.92. №A6. 5725-5734

Burlaga, L.F., Fitzenreiter, R., Lepping, R., Ogilvie, K., Szabo, A., Lazarus, A., Steinberg, J., Gloeckler, G., Howard, R., Michels, D., Farrugia, C., Lin, R.P., Larson, D.E. 1998. "A magnetic cloud containing prominence material: January 1997". *J. Geophys. Res*. v.103. №A1. 277-285

Burlaga, L.F., Szabo, A. 1999. "Fast and slow flows in the solar wind near the ecliptic at 1 AU?". *Space Sci. Rev*. v.87. 137-140

Burlaga, L.F., Lazarus, A. 2000. "Lognormal distributions and spectra of solar wind plasma fluctuations: Wind 1995-1998". *J. Geophys. Res*. v.105. №A2. 2357-2364

Burlaga, L.F. 2001. "Lognormal and multifractal distributions of the heliospheric magnetic field". *J. Geophys. Res*. v.106. №A8. 15917-15927

Celnikier, L.M., Harvey, C.C., Jegou, R., Kemp, M., Moricet, P. 1983. "A determination of the electron density fluctuation spectrum in the solar wind, using the ISEE propagation experiment". *Astron. Astrophys*. v.126. 293-298

Crooker, N.U., Shodhan, S., Gosling, J.T., Simmerer, J., Lepping, R.P., Steinberg, J.T., Kahler, S.W. 2000. "Density extremes in the solar wind". *Geophys. Res. Lett*. v.27. №23. 3769-3772

Dal Lago, A., Gonzalez, W.D., de Gonzalez, A.L.C., Vieira, L.E.A. 2001. "Compression of magnetic clouds in interplanetary space and increase in their geoeffectiveness". *J. Atmosph. Solar Terrest. Phys*. v.63. 451-455

Deeming, T.J. 1975. "Fourier Analysis with unequally-spaced data". *Astrophys. Space Sci*. v.36. 137-158

Denskat, K.V., Neubauer, F.M. 1982. "Statistical properties of low-frequency magnetic field fluctuations in the solar wind from 0.29 to 1 AU during solar minimum conditions: Helios-1 and Helios-2". *J. Geophys. Res*. v.A87. №4. 2215-2223

Feynman, J., Ruzmaikin, A. 1994. "Distributions of the interplanetary magnetic field revisited". *J. Geophys.Res*. v.99. №A9. 17645-17651

Freeman, J., Lopez, R. 1985 "The cold solar wind". *J. Geophys. Res*. v.90. №A10. 9885-9894

Grappin, R., Velli, M., Mangeney, A. 1991. "“Alfvénic” versus “standard” turbulence in the solar wind". *Ann. Geophys*. v.9. 416-426

Hartlep, T., Matthaeus, W.H., Padhye, N.S., Smith, C.W. 2000. "Magnetic field strength distribution in interplanetary turbulence". *J. Geophys. Res*. v.105. №A3. 5135-5139

Janardhan, P., Fujiki, K., Sawant, H.S., Kojima, M., Hakamada, K., Krishnan, R. 2008. "Source regions of solar wind disappearance events". *J. Geophys. Res*., v.113: A03102. https://doi.org/10.1029/2007JA012608

Kolmogorov, A.N. 1941. "The local structure of turbulence in incompressible viscous fluid for very large Reynolds' numbers". *Dokl. Akad. Nauk SSSR*. v.30. 301-305

Lazarus, A.J. 2000. "The day the solar wind almost disappeared". *Science*. v.287. №5461. 2172-2173

Lopez, R., Freeman, J. 1986. "Solar wind proton temperature - velocity relationship". *J. Geophys. Res*. v.91. №A2. 1701-1705

Lopez, R. 1987. "Solar cycle invariance in solar wind proton temperature relationships". *J. Geophys. Res*. v.92. №A10. 11189-11194

Luhmann, J., Zhang, T.-L., Petrinec, S.M., Russell, C.T., Gazis, P., Barnes, A. 1993. "Solar cycle 21 effects on the interplanetary magnetic field and related parameters at 0.7 and 1.0 AU". *J. Geophys. Res*. v.98. №A4. 5559-5572

Mangeney, A., Grappin, R., Velli, M. 1991. "Magnetohydrodynamic turbulence in the solar wind". In *Advances in Solar System Magnetohydrodynamics*, edited by E.R. Priest, A.W. Hood, 327-356. Cambridge: Cambridge University Press

Marsch, E., Muhlhauser, K.-H., Rosenbauer, H., Schwenn, R., FM. Neubauer, F.M. 1982. "Solar wind helium ions: observations of the Helios solar probes between 0.3 and 1 AU". *J. Geophys. Res*. v.87. 35-51

Marsch, E., Muhlhauser, K.H., Schwenn, R., Rosenbauer, H., Pilipp, W., Neubauer, F.M. 1982. "Solar wind protons: three-dimensional velocity distributions and derived plasma parameters measured between 0.3 and 1 AU". *J. Geophys. Res.* v.A87. №1. 52-72

Matthaeus, W.H., Goldstein, M.L., Roberts, D.A. 1990. "Evidence for the presence of quasi-two-dimensional nearly incompressible fluctuations in the solar wind". *J. Geophys. Res.* v.95. 20673-20683

McComas, D.J., Elliott, H.A., Schwadron, N.A., Gosling, J.T., Skoug, R.M., Goldstein, B. E. 2003. "The three-dimensional solar wind around solar maximum". *Geophys. Res. Lett.* v.30(10). 1517-1521

Newbury, J.A., Russell, C.T., Phillips, J.L., Gary, S.P. 1998. "Electron temperature in the ambient solar wind: typical properties and a lower bound at 1 AU". *J. Geophys. Res.* v.103. 9553-9566

Ogilvie, K., Coplan, M. Zwickl, R. 1982. "Helium, hydrogen, and oxygen velocities observed on ISEE 3". *J. Geophys. Res.* v.87. №A9. 7363-7369

Owens, M.J., Cargill, P.J. 2002. "Correlation of magnetic field intensities and solar wind speeds of events observed by ACE". *J. Geophys. Res.* v.107, №A5: 1050. https://doi.org/10.1029/2001JA000238

Owens, M.J., Cargill, P.J., Pagel, G.L., Siscoe, G.L., Crooker, N.U. 2005. "Characteristic magnetic field and speed properties of interplanetary coronal mass ejections and their sheath regions". *J. Geophys. Res.* v.110: A01105. https://doi.org/10.1029/2004JA0010814

Perri, S., Balogh, A. 2010. "Differences in solar wind cross-helicity and residual energy during the last two solar minima". *Geophys. Res. Lett.* v.37. 17102-17105

Pilipp, W.G., Schwenn, R., Marsch, E., Muhlhauser, K.-H., Rosenbauer, H. 1981. "Electron characteristics in the solar wind as deduced from Helios observations". In *Solar Wind Four*, edited by H. Rosenbauer, 241-249. Garching: Report No MPAE-W-100-81-31, Max-

Plank-Institute fur Aeronomie, Katlenburg-Lindau and Max-Plank-Institute fur extraterrestrische Physik

Pilipp, W.G., Miggenrieder, H., Montgomery, M.D., Muhlhauser, K.-H., Rosenbauer, H., Schwenn, R. 1987. "Characteristics of electron velocity distribution functions in the solar wind derived from the Helios plasma experiment". *J. Geophys. Res*. v.92. 1075-1092

Rosenbauer, H., Schwenn, R., Marsch, E., Meyer, B., Miggenrider, H., Montgomery, M.D., Muhlhauser, K.-H., Pilipp, W.G., Voges, W., Zink, S.M. 1977. "A survey of initial results of the Helios plasma experiment". *Journal of Geophysics - Zeitschrift fur Geophysik*. v.42. 561-580

Richardson, I.G., Cane, H.V. 1995. "Regions of abnormally low proton temperature in the solar wind (1965–1991) and their association with ejecta". *J. Geophys. Res*. v.100. №A12. 23397-23412

Skoug, R.M., Gosling, J.T., Steinberg, J.T., McComas, D.J., Smith, C.W., Ness, N.F., Hu, Q., Burlaga, L.F. 2004. "Extremely high speed solar wind: 29–30 October 2003". *J. Geophys. Res*. v.109: A09102. https://doi.org/10.1029/2004JA010494

Smith, E.J., Neugebauer, M., Balogh, A., Bame, S.J., Erdos, G., Forsyth, R.J., Goldstein, B.E., Phillips, J.T., Tsurutani, B.T. 1993. "Disappearance of the heliospheric sector structure at Ulysses". *Geophys. Res. Lett*. v.20. 2327-2330

Sorriso-Valvo, L., Marino, R., Carbone, V., Noullez, A., Lepreti, F., Veltri, P., Bruno, R., Bavassano, B., Pietropaolo, E. 2007. "Observation of inertial energy cascade in interplanetary space plasma". *Phys. Rev. Lett*. v.99(11). 115001-115001

Spreiter, J.R., Summers, A.L., Alksne, A.Y. 1966. "Hydromagnetic flow around the magnetosphere". *Planet. Space Sci*. v.14. 223-250

Schwenn, R., Muhlhauser, K.H., Marsch, E., Rosenbauer, H. 1981. "Two states of the solar wind at the time of solar activity minimum. 2. Radial gradients of plasma parameters in fast and slow streams". In *Solar*

Wind Four, edited by H. Rosenbauer, 126-130. Garshing: Report No MPAE-W-100-81-31, Max-Plank-Institute fur Aeronomie, Katlenburg-Lindau and Max-Plank-Institute fur extraterrestrische Physik

Taylor, G.I. 1938. "The spectrum of turbulence". *Proc. Roy. Soc.* v.A164. 476-490

Tsurutani, B., Gonzalez, W. 1987. "The cause of high-intensity long-duration continuous AE activity (HILDCAAS): interplanetary Alfven wave trains". *Planet. Space Sci.* v.35. №4. 405-412

Tsurutani, B.T., Gonzalez, W.D., Gonzalez, A.L.C., Tang, F., Arballo, J.K., Okada, M. 1995. "Interplanetary origin of geomagnetic activity in the declining phase of the solar cycle". *J. Geophys. Res.* v.100. №A11. 21717-21733

Usmanov, A.V., Goldstein, M.L., Ogilvie, K.W., Farrell, W.M., Lawrence, G.R. 2005. "Low-density anomalies and sub-Alfvenic solar wind". *J.Geophys.Res.*v.110:A01106.
https://doi.org/10.1029/2004JA010699

Veselovsky, I.S., Dmitriev, A.V., Panasenko, O.A., Suvorova, A.V. 1999. "Solar cycles in the energy and mass outputs of the heliospheric plasma". *Astronomy Reports*. v.3. 485-486

Veselovsky, I.S., Dmitriev, A.V., Suvorova, A.V. 2010. "Algebra and statistics of the solar wind". *Cosmic Research*. v.48. №2. 113-128

Wicks, R.T., Horbury, T.S., Chen, C.H.K., Schekochihin, A.A. 2010. "Power and spectral index anisotropy of the entire inertial range of turbulence in the fast solar wind". *MNRAS*. v.407. L31-L35

Zurbuchen, T.H., Hefti, S., Fisk, L.A., Gloeckler, G., Schwadron, N.A., Smith, C.W., Ness, N.F., Skoug, R.M., McComas, D.J., Burlaga, L.F. 2001. "On the origin of microscale magnetic holes in the solar wind". *J. Geophys. Res.* v.106. №A8. 16001-16010

References to Chapter 4

Арцимович, Л.А. 1978. *Атомная физика и физика плазмы*. Москва: Наука. (in Russian)

Ландау, Л.Д., Лифшиц, Е.М. 1992. *Электродинамика сплошных сред*. Москва: Наука. (in Russian)

Лифшиц Е.М., Питаевский Л.П. 1979. *Физическая кинетика*. Москва: Наука. (in Russian)

Лойцянский, Л.Г. 1973. *Механика жидкости и газа*. Москва: Наука. (in Russian)

Несмянович, А.Т. 1971. "О радиальном характере корональных лучей". *Солнечные данные*. №9. 74-76 (in Russian)

Пикельнер, С.Б. 1950. "К теории солнечной короны". *Известия КрАО*. №5. 34-58 (in Russian)

Пикельнер, С.Б. 1950. "Диссипация короны и её значение". *Доклады АН СССР*. т.72. 255-258 (in Russian)

Писанко, Ю.В. 1985. "Расчёт стационарных магнитогидродинамических течений в солнечной короне". *Геомагнетизм и аэрономия*. т.25. №1. 17-22 (in Russian)

Писанко, Ю.В. 1985. "Пример численного расчёта пространственной структуры солнечного ветра". *Геомагнетизм и аэрономия*. т.25. №3. 371-375 (in Russian)

Писанко, Ю.В. 1986. "Токовая система солнечного ветра: результаты численного расчёта". *Геомагнетизм и аэрономия*. т.26. №1. 1-5 (in Russian)

Писанко, Ю.В. 1989. "Об одном подходе к расчёту доальвеновского солнечного ветра". в кн: *Математические Модели Ближнего Космоса*, под редакцией И.И. Алексеев, И.С. Веселовский, А.П. Кропоткин, 5-8. Москва: МГУ, ISSN-5-211-01920-2 (in Russian)

Уиндхэм, Д.Д. 1968. "Радиоастрономические наблюдения, связанные с проблемой солнечного ветра". в кн: *Солнечный Ветер*, под редакцией Р.Д. Маккин, М. Нейгебауер, 137-150. Москва: Мир (in Russian)

Чашей, И.В. 1992. "Корона и солнечный ветер в квазидипольном магнитном поле". *Астрономический журнал*. т.69. №1. 192-200 (in Russian)

Шабанский, В.П., Шистер, А.Р. 1973. "Сверхзвуковое истечение плазмы из вращающегося источника". *Доклады АН СССР*. т.211. 825-828 (in Russian)

Шабанский, В.П., Шистер, А.Р. 1973. "Взаимодействие потоков с разными скоростями в солнечном ветре". *Геомагнетизм и аэрономия*. т.13. №4. 569-573 (in Russian)

Alfven, H. 1977. "Electric currents in cosmic plasmas". *Rev. Geophys. Space Phys*. v.15. №3. 271-284

Altschuler, M.D., Newkirk G., Jr. 1969. "Magnetic fields and the structure of the solar corona. I. Methods of calculating coronal fields". *Solar Phys*. v.9. №1. 131-149

Aly, J.J. 1984. "On some properties of force-free magnetic fields in infinite regions of space". *Astrophys. J*. v.283. 349-362

Aly, J.J. 1991. "How much energy can be stored in a three-dimensional force-free magnetic field?". *Astrophys. J*. v.375. L61-L64

Balogh, A., Smith, E.J., Tsurutani, B.T., Southwood, D.J., Forsyth, R.J., Hornbury, T.S. 1995. "The heliospheric magnetic field over the south polar region of the sun". *Science*. v.268. 1007-1010

Bondi, H. 1952. "On spherically symmetric accretion". *MNRAS*. v.112. 195-204

Burlaga, L. 1988. "Magnetic clouds and force-free fields with constant alpha". *J. Geophys.Res*. v.93. 7217-7224

Chen, J. 1989. "Effects of toroidal forces in current loops embedded in a background plasma". *Astrophys. J.* v.338. 453-470

Coles, W.A. Esser, R., Lovhaug, U.-P., Markkanen, J. 1991. "Comparison of solar wind velocity measurements with a theoretical acceleration model". *J. Geophys. Res.* v.96. 13849-13859

Decker, R.B., Roelof, E.C., Krimigis, S.M., Hill, M.E. 2006. "Low-energy ions near the termination shock". In *Physics of the Inner Heliosheath: Voyager Observations, Theory and Future Prospects*, edited by J. Heerikhuisen, V. Florinsky, G.P. Zank, N.V. Pogorelov, 73-78. Melville, NY: AIP

Delannee, C., Delaboudiniere, J.-P., Lamy P. 2000. "Observation of the origin of CMEs in the low corona". *Astron. Astrophys*. v.355. 725-742

Dryer, M. 1996. "Comments on the origin of coronal mass ejections". *Sol. Phys*. v.169. 421-429

Esser R., Edgar, R.J. 2000. "Reconciling spectroscopic electron temperature measurements in the solar corona with in situ charge states observations". *Astrophys. J.* v.532. L71-L74

Feldman, W.C., Asbridge, J.R., Bame, S.J., Gosling, J.T. 1977. "Plasma and magnetic fields from the Sun". In *The Solar Output and its Variation*, edited by D.R. White, 351-382. Boulder: Colorado Associated University Press

Gibson, S.E., Biesecker, D., Guhathakurta, M., Hoeksema, J.T., Lazarus, A., Linker, J., Mikic, Z., Pisanko, Y., Riley, P., Steinberg, J., Strachan, L., Szabo, A., Thompson, B., Zhao, X.P. 1999. "The three-dimensional coronal magnetic field during Whole Sun Month". *Astrophys. J.* v.520. №2. 871-879

Gloeckler, G., Fisk, L.A., Zurbuchen, T.H., Schwadron, N.A. 2000. "Sources, injection and acceleration of heliospheric ion populations". In *Acceleration and Transport of Energetic Particles Observed in the*

Heliosphere, edited by R.A. Mewaldt, M. Miller, J.R. Jokipij, M.A. Lee, T.H. Zurbuchen, E. Mobius, 221-228. Melville, NY: AIP

Gloeckler, G. 2003. "Ubiquitous suprathermal tails on the solar wind and pickup ion distributions". In *Solar Wind Ten*, edited by M. Velli, R. Bruno, F. Malara, 583-588. Mellvile, NY: AIP

Grigor'ev, V.I., Grigor'eva, E.V., Pisanko, Yu.V., Rostovsky, V.S. 2005. *Baro-electric Effect and Celestial Magnetism*. Moscow: KomKniga

Hansen, S.F., Sawyer, C., Hansen, R.T. 1974. "K-corona and magnetic sector boundaries". *Geophys. Res. Lett*. v.1. 13-15

Howard, R.A., Koomen, M.J. 1974. "Observation of sectored structure in the outer solar corona: correlation with interplanetary magnetic field". *Solar Phys*. v.37. №2. 469-475

Hugoniot, P.-H. 1886. "Sur un theoreme relatif au mouvement peramnent et a l'ecoulement des fluids". *Comptes Rendus de Seances de l'Academie des Sciences*. v.103. 1178-1181

Van de Hulst, H.C. 1953. "The chromosphere and the corona". In *The Sun*, edited by G.P. Kuiper, 207-321. Chicago: University of Chicago Press

Hundhausen, A.J. 1977. "An interplanetary view of coronal holes". In *Coronal Holes and High Speed Wind Streams*, edited by J.B. Zirker, 226-329. Boulder: Colorado associated university press

Hundhausen, A.J., Sawyer, C.B., House L., Illing R.M.E., Wagner W.J. 1984. "Coronal mass ejections observed during the solar maximum mission: latitude distribution and rate occurrence". *J. Geophys. Res*. v.89. 2639-2646

Jokers, K. 1970. "Solar wind models based on exospheric theory". *Astron. Astrophys*. v.6. 219-239

Klimchuk, J.A. 1990. "Shear-induced inflation of coronal magnetic fields". *Astrophys. J*. v.354. 745-754

Kohl, J.L., Noci, G., Antonucci, E., Tondello, G., Huber, M.C.E., Cranmer, S.R., Strachan, L., Panasyuk, A.V., Gardner, L.D., Romoli, M., Fineschi, S., Dobrzycka, D., Raymond, J.C., Nicolosi, P., Siegmund, O.H.W., Sparado, D., Benna, C., Ciaravella, A., Giordano, S., Habbal, S.R., Karovska, M., Li, X., Martin, R., Michels, J.D., Modigliani, A., Naletto, G., O'Neal, R.H., Pernechele, C., Poletto, G., Smith, P.L., Suleiman, R.M. 1998. "UVCS/SOHO empirical determinations of anisotropic distributions in the solar corona". *Astrophys. J.* v.501. №1. L127-L131

Kuznetsov, V.D., Hood, A.W. 1997. "Non-equilibrium of magnetic flux tubes emerging into the solar corona". *Sol. Phys*. v.171. 61-80

Kuznetsov, V.D., Hood, A.W. 2000. "A phenomenological model of coronal mass ejection". *Adv. Space Res*. v.26. 539-542

Maksimovich, M., Pierrard, V., Lemaire, J.F. 1997. "A kinetic model of the solar wind with Kappa distribution functions in the corona". *Astron. Astrophys*. v.324. 725-734

McAllister, A.H., Hundhausen, A.J. 1996. "The relation of Yohkoh coronal arcade events to coronal streamers and CMEs". In *Solar Drivers of Interplanetary and Terrestrial Disturbances*, edited by K.S. Balasubramaniam, S.L. Keil, R.N. Smart, 171-184. San Francisco: Astron. Soc. of the Pacific

Moussas, X., Polygiannakis, J., Kakouris, A., Alevizos, A. 1996. "The 26-day variation of energetic particles at high heliographic latitudes and corotating vortex structures". In *Solar and Astrophysical Magnetohydrodynamic Flows*, edited by K. Tsinganos, 46. Heraclion: NATO Advanced Study Institute

Mullan, D.J. 1990. "Sources of the solar wind: what are the smallest-scale structures?". *Astron. Astrophys*. v.232. 520-535

Mulligan, T., Russell, C.T., Luhmann, J.G. 1998. "Solar cycle evolution of the structure of magnetic clouds in the inner heliosphere". *Geophys. Res. Lett*. v.25. 2959-2962

Munro, R.H., Jackson, B.V. 1977. "Physical properties of a polar coronal hole from 2 to 5 solar radii". *Astrophys. J.* v.213. №3. 874-886

Neslusan, L. 2001. "On the global electrostatic charge of stars". *Astron. Astrophys*. v372. 913-915

Pannekoek, A. 1922. "Ionization in stellar atmospheres". *Bull. Astron. Inst. Neth.* v.1. 107-118

Parker, E.N. 1958. "Dynamics of the interplanetary gas and magnetic fields". *Astrophys. J.* v.128. 644-675

Parker, E.N. 1960. "The hydrodynamic theory of solar corpuscular radiation and stellar winds". *Astrophys. J.* v.132. 821-866

Parker, E.N. 1983. "Magnetic neutral sheets in evolving fields. II Formation of the solar corona". *Astrophys. J.* v.264. 642-647

Pisanko, Yu.V. 1994. "3-D subalfvenic transonic solar wind in the flow formation region: results of numerical MHD simulations". In *Solar-terrestrial Energy Program*, edited by D.N. Baker, V.O. Papitashvili, M.J. Teague, 791-794. Oxford, New York, Seoul, Tokyo: Pergamon Press

Pisanko, Yu.V. 1996. "The global solar magnetic field as a "controller" for the origin and acceleration model of solar wind streams". *Adv. Space Res*. v.17. №3. (3)61-(3)64

Pisanko, Yu.V. 1997. "The polar solar wind: a linear, force-free field, 3-D MHD model". *Solar Phys*. v.172. 345-352

Pisanko, Yu.V. 1997. "The polar solar wind: an additional acceleration due to the rotation of the magnetic sun". In *The Corona and Solar Wind Near Minimum Activity*, edited by O. Kjeldseth-Moe, A. Wilson, 605-608. ESA SP-404

Pisanko, Yu.V. 1998. "Electrodynamics of the polar solar wind additional acceleration". In *A Crossroads for European Solar and Heliospheric Physics*, edited by E.R. Priest, F. Moreno-Insertis, R.A. Harris, 297-300. ESA SP-417

Pisanko, Yu.V., Yakovlev, O.I. 2020. “Plasma polarization electric field derived from radio sounding of solar wind acceleration region with spacecraft signals”. *Adv. Space Res.* v.65. 1048-1053

Pneuman, G.W. 1983. “Ejection of magnetic fields from the sun: acceleration of a solar wind containing diamagnetic plasmoids”. *Astrophys. J.* v.265. №1. 468-482

Pneuman, G.W., Cargill, P. 1985. “Evolution of diamagnetic material in a nonuniform magnetic field”. *Astrophys. J.* v.288. 653-664

Potemra, T.A. 1986. “Auroral field-aligned currents: APL’s contribution to their discovery”. *Johns Hopkins APL Tech. Dig.* v.7. 397-402

Rosseland, S. 1924. “Electrical state of a star”. *MNRAS.* v.84. №9. 720-729

Schatten, K.H. 1971. “Current sheet magnetic model for the solar corona”. *Cosmic Electrodyn.* v.2. 232-246

Schluter A. 1957. “Solar radio emission and the acceleration of magnetic storm particles”. In *IAU Symposium №4 on Radio Astronomy*, edited by H.C. van de Hulst, 356-357. Cambridge: Cambridge University Press

Schwenn, R., Muhlhauser, K.H., Marsch, E., Rosenbauer H. 1981. “Two states of the solar wind at the time of solar activity minimum. 2. Radial gradients of plasma parameters in fast and slow streams”. In *Solar Wind Four*, edited by H. Rosenbauer, 126-130. Garching: Report No MPAE-W-100-81-31, Max-Plank-Institute fur Aeronomie, Katlenburg-Lindau and Max-Plank-Institute fur extraterrestrische Physik

Scrauner, J.A., Scherrer, P.H. 1994. “East-west inclination of large-scale photospheric magnetic fields”. *Solar Phys.* v.153. 131-141

Shoub, E.C. 1983. “Invalidity of local thermodynamic equilibrium for electrons in the solar transition region”. *Astrophys. J.* v.266. 339-369

Simunac, K.D.C., Armstrong, T.P. 2004. "Solar cycle variations in solar and interplanetary ions observed with Interplanetary Monitoring Platform 8". *J. Geophys. Res*. v.109: A10101. https://doi.org/10.1029/2003JA010194

Stegun, I.A., 1972. "Miscellaneous functions". In *Handbook of Mathematical Functions with Formulas, Graphs and Mathematical Tables*, edited by M. Abramowitz, I.A. Stegun, 998-1010. National Bureau of Standards, Applied Mathematics Series 55

Steinolfson, R.S., Suess, S.T., Wu, S.T. 1982. "The steady global corona". *Astrophys. J.* v.255. №2. 730-742

Svalgaard, L., Wilcox, J.M., Duvall, T.L., Jr. 1974. "A model combining the polar and the sector structured solar magnetic fields". *Solar Phys*. v.37. №1. 157-172

Tousey, R. 1973. "The solar corona". In *Space Research XIII*, edited by M.J. Rycroft, S.K. Runcorn, 713-730. Berlin: Academic-Verlag

Usmanov, A.V. 1993. "A global numerical 3-D MHD model of the solar wind". *Solar Phys*. v.146. 377-396

Usmanov, A.V. 1996. "A global 3-D MHD solar wind model with Alfven waves". In *Solar Wind Eight*, edited by D. Winterhalter, 141-144. Woodbury: AIP press

Vsekhsvjatsky, S.K. 1963. "The structure of the solar corona". In *The Solar Corona*, edited by J.M. Evans, 129-136. New York and London: Academic press

Weber, E.J., Davis L., Jr. 1967. "The angular momentum of the solar wind". *Astrophys. J.* v.148. 217-227

Woo, R., Armstrong, J.W. 1992. "Observations of large-scale structure in the inner heliosphere with Doppler scintillation measurements". In *Solar Wind Seven*, edited by E. Marsch, R. Schwenn, 319-322. New York, London, Seoul, Tokyo: Pergamon Press

Woo, R., Habbal, S.R. 1999. "Imprint of the sun on the solar wind". *Astrophys. J.* v.510. №1. L69-L72

Yakovlev, O.I., Pisanko, Yu.V. 2018. "Radio sounding of the solar wind acceleration region with spacecraft signals". *Adv. Space Res.* v.61. 552-566

Yakovlev, O.I., Yakovlev, Y.O. 2019. "Analysis of radial dependencies of the electron concentration and plasma velocity in the region of accelerated solar wind". *Cosmic Research*. v.57(4). 237-242

References to Chapter 5

Антонова, А.Е., Шабанский, В.П. 1968. "О структуре геомагнитного поля на больших расстояниях от Земли". *Геомагнетизм и аэрономия*. т.8. 801-811 (in Russian)

Арцимович, Л.А. 1966. *Элементарная физика плазмы*. Москва: Атомиздат. (in Russian)

Веселовский, И.С. 1984. *Физика межпланетной плазмы. Итоги науки и техники. Серия «Исследование космического пространства»*, т.22. Москва: ВИНИТИ. (in Russian)

Веселовский, И.С. 2008. "Происхождение солнечного ветра: астрофизический и плазменно-физический аспекты проблемы". *Солнечно-земная физика*. т.12. №1. 93-98 (in Russian)

Веселовский, И.С. 2017. "Переменный солнечный ветер". *Астрономический вестник*. т.51. №3. 269-272 (in Russian)

Гибсон, Э. 1977. *Спокойное Солнце* (перевод с английского под редакцией Э.В. Кононовича). Москва: Мир. (in Russian)

Ефимов, А.И., Чашей, И.В., Шишов, В.И., Берд, М.К. 1993. "О флуктуациях фарадеевского вращения при радиопросвечивании околоземной плазмы". *Письма в Астрономический журнал*. т.19. №2. 143-152 (in Russian)

Чашей, И.В., Шишов, В.И. 1987. "Формирование потоков энергии и массы солнечного ветра в модели с волновым источником". *Геомагнетизм и аэрономия*. т.27. №5. 705-711 (in Russian)

Чашей, И.В., Шишов, В.И. 1988. "Самосогласованная модель спокойной короны с волновым источником энергии". *Астрономический журнал*. т.65. 157-166 (in Russian)

Alfven, H. 1947. "Granulation, magnetic-hydrodynamic waves, and the heating of the solar corona". *MNRAS*. v.107. 211-219

Biermann, L. 1948. "Uber die ursache der chromospharischen turbulenz und des UV-exzesses der sonnenstrahlung", *Zeitschrift fur Astrophysik*. v.25. 161-177

Boltenkov, B.S., Gartmanov, V.N., Kocharov, G.E., Mamyrin, B.A., Naidenov, V.O. 1972. "Measurement of the isotopic composition of particle flux carried out on spacecraft Soyuz, Zond 8 and Luna 16". *Space Res*. v.12. 1487-1491

Bondi, H. 1952. "On spherically symmetric accretion". *MNRAS*. v.112: 195-204

Bruckner, K.A., Jorna, S. 1973. *Laser Driven Fusion*. Ann Arbor, Michigan: KMS Fusion Inc.

Cox, D. Tucker, W. 1969. "Ionization equilibrium and radiative cooling in a low-density plasma". *Astrophys. J*. v.157. 1157-1167

Feldman, W.C., Asbridge, J.R., Bame, S.J., Gosling, J.T. 1977. "Plasma and magnetic fields from the Sun". In *The Solar Output and Its Variation*, edited by D.R. White, 351-382. Boulder: Colorado Associated University Press

Fox, N.J., Velli, M.A., Bale, S.D., Decker, R., Driesman, A., Howard, R.A., Kaspar, J.C., Kinnison, J., Kusterer, M., Lario, D., Lockwood, M.K., McComas, D.J., Raouafi, N.E., Szabo, A. 2016. "The Solar Probe Plus mission: humanity's first visit to our star". *Space Sci. Rev*. v.204. 7-48.

Geiss, J., Hirt, P., Leutwyler, H. 1970. "On acceleration and motion of ions in corona and solar wind". *Solar Phys*. v.12. 458-483

Geiss, J., Gloecker, G. 2003. "Isotopic composition of H, He and Ne in the protosolar cloud". *Space Sci. Rev*. v.106. 3-18

Gibson, S.E., Biesecker, D., Guhathakurta, M., Hoeksema, J.T., Lazarus, A., Linker, J., Mikic, Z., Pisanko, Y., Riley, P., Steinberg, J., Strachan, L., Szabo, A., Thompson, B., Zhao, X.P. 1999. "The three-dimensional coronal magnetic field during Whole Sun Month". *Astrophys. J*. v.520. №2. 871-879

Gingerich, O., Noyes, R.W., Kalkofen, W., Cury, Y. 1971. "The Harvard-Smithsonian reference atmosphere". *Solar Phys*. v.18. 347-365

Hester, J., Desch, S.J., Healy, K.R., Leshin, L.A. 2004. "The gradle of the solar system". *Science*. v.304. 1116-1117

Jacques, S.A. 1977. "Momentum and energy transport by waves in the solar atmosphere and solar wind". *Astrophys. J*. v.215. 942-951

Kallenbach, R. 2003. "Isotopic fractionation by plasma processes". *Space Sci. Rev*. v.106. 305-316

Korevaar, P. 1989. "Time-dependent corona models: coronae with accretion". *Astron. Astrophys*. v.226. 209-214

Kuperus, M. 1969. "The heating of the solar corona". *Space Sci. Rev*. v.9. 713-739

McKee, C.F., Ostriker, E.C. 2007. "Theory of star formation". *Annual Review of Astronomy & Astrophysics*. v.45. №1. 565-687

Michalas, D. 1978. *Stellar Atmospheres*. San Francisco: W.H. Freeman and Co.

Neugebauer, M., Snyder, C.W. 1966. "Mariner 2 observations of the solar wind, 1. Average properties". *J. Geophys. Res*. v.71. 4469-4484

Osterbrock, D.E. 1959. “Heating of the solar chromosphere, plages, and corona by magneto-hydrodynamic waves”. *Astrophys. J.* v.134. 347-388

Parker, E.N. 1958. “Dynamics of interplanetary gas and magnetic fields”. *Astrophys. J.* v.128. 664-676

Pisanko, Yu.V. 2005. “A model of the thermal expansion of the solar corona with allowance for radiation”. *Solar System Research.* v.39. №5. 426-430

Pomoell, J., Poedts, S. 2018. “EUHFORIA: European heliospheric forecasting information asset”. *J. Space Weather Space Clim.* v.8: A35. https://doi.org/10.1051/swsc/2018020

Schatzman, E. 1949. “The heating of the solar corona and chromosphere”. *Ann. Astrophys*. v.12. 203-228

Schwarzschield, M. 1948. “On noise arising from the solar granulation”. *Astrophys. J.* v.107. 1-5

Shabansky, V.P. 1965. “On the first phase of a magnetic storm”. *Space Res*. v.5. 125-147

Triskova, L., Veselovsky, I.S. 1992. “On the large-scale magnetic field structure in the outer heliosphere”. In *Solar Wind Seven*, edited by E. Marsch, R. Schwenn, 297-300. New York, London, Seoul, Tokyo: Pergamon Press

Velli, M. 1994. “From supersonic winds to accretion: comments on the stability of stellar winds and related flows”. *Astrophys. J.* v.432. L55-L58

Veselovsky, I.S. 1982. “On the screening of the interstellar electric and magnetic fields by the solar wind”. *Astrophys. Space Sci.* v.86. 209-213

Washimi, H. 1973. “Self-focusing of transverse waves in a magnetoplasma”. *J. Phys. Soc. Japan.* v.34. 1373-1383

Washimi, H., Karpman, V.I. 1976. "The ponderomotive force of a high-frequency electromagnetic field in a dispersive medium". *Soviet Phys., JETP*. v.44. 528-531

"*The Sun and Heliosphere in Three Dimensions*" – Report of the NASA Science Definition Team for the STEREO mission, 69p.